ESSENTIAL HEALTH AND SAFETY STUDY SKILLS

ESSENTIAL HEALTH AND SAFETY STUDY SKILLS

JONATHAN BACKHOUSE

LONDON AND NEW YORK

First edition published 2013
by Routledge
2 Park Square, Milton Park, Abingdon, Oxon OX14 4RN

Simultaneously published in the USA and Canada
by Routledge
711 Third Avenue, New York, NY 10017

Routledge is an imprint of the Taylor & Francis Group, an informa business

British Library Cataloguing in Publication Data
A catalogue record for this book is available from the British Library

Library of Congress Cataloging in Publication Data
Backhouse, Jonathan, author.
Essential health and safety study skills / Jonathan Backhouse.
pages cm
Includes bibliographical references and index.
1. Safety education, Industrial—Great Britain—Examinations—Study guides. 2. Industrial hygiene—Great Britain—Examinations—Study guides. I. Title.
T55.2.B33 2013
658.3'82—dc23 2012042517

ISBN: 978-0-415-62909-6 (pbk)
ISBN: 978-0-203-10020-2 (ebk)

Typeset in Melior and Univers
by Cenveo Publisher Services

Printed and bound in Great Britain by MPG Printgroup

CONTENTS

vi

ABOUT THE AUTHOR

Jonathan Backhouse is an occupational safety and health practitioner and trainer, with over ten years' experience in consultancy, training and course development in the UK and overseas. He is an executive member of IOSH Tees Branch and NEBOSH examiner.

Jonathan gained occupational experience by spending four years in engineering management where he implemented an integrated Safety, Health and Environmental Management System.

Professional qualifications include Masters in Professional Practice in the Context of Education, BA (Hons) in Education and Training; NEBOSH National Diploma in Occupational Health and Safety, and also in Environmental Management; and Certificate in Education.

FOREWORD

It seems like only five minutes since I was studying for my own first health and safety qualification. It's actually getting on for 30 years, and was one of the best decisions I have ever made. As a mentor of mine recommended a career in health and safety to me, I'm pleased and actually quite proud to recommend this worthy and rewarding career to you. The bus is definitely heading towards 'greater competency in the health and safety profession', so that's the journey for you to contemplate; and, when you're ready, the bus to board. But, having picked up this book, you've probably decided that already.

The number of people in the health and safety profession with higher qualifications has grown since the first courses offered by Aston and Loughborough universities in the early 1990s. There are more women in our profession now than ever before, and more people seem to be choosing health and safety as their first career. Organizations and their stakeholders demand much safer, healthier operations these days. Quite recently, 13 major construction projects were completed on London's Olympic Park without a single fatality and with world-class (low) incident rates. Others will be compelled to follow.

Of course, back in the old days, anyone could be a health and safety officer. For some, it was the boot of redundancy; for others the safety job. These days, things have changed. The IOSH has over 40,000 members in its six membership categories, and it is now the world's largest health and safety organization. Its strategic plan states that it will have 60,000 members by 2017 by growing internationally as well as at home.

All this gives rise to opportunities for you if you can meet the competency challenge. For the right candidates, there are better jobs, better prospects and better salaries than ever before; but first, you need to be qualified. That's where this book comes in. Jonathan Backhouse knows health and safety qualifications extremely well. As a qualified instructor, he has taught dozens of courses to hundreds of candidates, and, as an examiner, he has marked hundreds of papers. If anyone knows which study skills to share with you, it's Jonathan.

I am writing this Foreword towards the end of a wonderful summer of Olympic and Paralympic sport. It has created a wonderful memory for all of us, and leaves Britain with a permanent and positive legacy. As you commence your programme of study – whether a basic health and safety award or a professional certificate qualification (such as the NEBOSH General Certificate) – I wonder what your legacy might become. Will you save colleagues' lives? Reduce ill-health and suffering? Or just gain great satisfaction from your work protecting others? Years ago, someone told me that if I chose my career wisely, I'd never work a day in my life. I hope you enjoy your career in health and safety as much as I have. Good luck with your studies, and in your exams.

Stephen Asbury, MBA FRSA CEnv CFIOSH
IOSH Council 1998 to 2012
Chair, IOSH Professional Committee / Professional Ethics Committee
1995 to 2001 and 2004 to date
Managing Director, Corporate Risk Systems Limited 2001 to date

ACKNOWLEDGEMENTS

This book would not have been possible without the support, and encouragement from my wife Diane. Thank you for believing I could do it!

I would also like to thank:

- Stephen Asbury, who first deposited the idea of me writing a book some two or three years ago, and who has kindly written the Foreword.
- Kara Milne, my publisher, who made my dream a reality.
- Dr Ian Smythe, for giving permission to use his Adult Dyslexia Checklist.

CLARIFICATION OF TERMS USED

For clarity and consistency, the following terms will be used throughout this book:

Accredited course	Accredited and credit rated by the Office of Qualifications and Examinations Regulation: Ofqual. OR Accredited and credit rated by the Scottish Qualifications Authority (SQA).
Awarding body	Professional body that has developed accredited courses for health and safety.
Course provider	A training provider/centre that is able to deliver accredited courses on behalf of an awarding body.
Trainer	Trainer/tutor/teacher/lecturer who has been authorized to deliver health and safety courses by the course provider or awarding body. They may or may not hold a teaching/training qualification.
Professional bodies	IOSH – Institution of Occupational Safety and Health. IIRSM – International Institute of Risk and Safety Management.
Professional certificate qualifications	Certificate qualifications that will lead to professional membership with IOSH or IIRSM or similar; for example, British Safety Council Certificates and many of the NEBOSH Certificates.

ACRONYMS USED

The following acronyms have been used in this book:

BDA	British Dyslexia Association
BSC	British Safety Council
CIEH	Chartered Institute of Environmental Health
HABC	Highfield Awarding Body of Compliance
HSE	Health and Safety Executive
IIRSM	International Institute of Risk and Safety Management
IOSH	Institution of Occupational Safety and Health
NEBOSH	National Examination Body of Occupational Safety and Health
PCQs	Professional Certificate Qualifications
RRR	Receive – Remember – Recall
RSPH	Royal Society for Public Health

INTRODUCTION

The driving force behind writing this book was the belief that if the author could, in 2000, pass the NEBOSH Certificate in Health and Safety, anyone can! At this time the author had no working experience of health and safety, had left school with two GCSEs at C or above, and had been diagnosed with dyslexia after leaving school.

This book has been specifically written for anyone who is undertaking a health and safety award or certificate-level qualification.The essential study skills will apply to all levels of health and safety qualifications, regardless of the awarding body.

The purpose of this book is to provide you with the essential study skills needed for classroom and distance/e-learning courses in health and safety.The focus is aimed at health and safety students studying health and safety professional certificate qualifications (for example, the NEBOSH General Certificate); however, these principles can also apply across the range of health and safety courses, from award to diploma-level qualifications provided by all awarding bodies.

This book will also provide you with essential examination skills for answering multiple-choice questions at award-level (1–3) qualifications and written exams for health and safety professional certificate qualifications and CIEH level 4 Award.

The case studies are real-life examples; for confidentiality, names and some minor details have been changed, with the exception of the authors.

Jonathan Backhouse
MA BA(Hons) DipNEBOSH EnvDipNEBOSH CertEd
FifL GradIOSH AIEMA

BENEFITS OF HEALTH AND SAFETY TRAINING

Health and safety, some would say, is not the most interesting of subjects to study. Health and safety often receives a bad press; for example, when headlines read 'Health and safety bans playing this, riding that ... But what is often missed is that health and safety saves lives, reduces serious accidents and prevents work-related ill-health.

The Health and Safety Executive (HSE) outlines the benefits of health and safety training.

Providing health and safety information and training helps you to:

- Ensure your employees are not injured or made ill by the work that they do.
- Develop a positive health and safety culture, where safe and healthy working becomes second nature to everyone.
- Find out how you could manage health and safety better.
- Meet your legal duty to protect the health and safety of your employees.

> The Health and Safety at Work etc. Act 1974 requires an employer to provide whatever information, instruction, training and supervision as is necessary to ensure, so far as is reasonably practicable, the health and safety at work of their employees.
>
> This is expanded by the Management of Health and Safety at Work Regulations 1999, which identify situations where health and safety training is particularly important; for example, when people start work, on exposure to new or increased risks and where existing skills may have become rusty or need updating.
>
> Source: http://www.hse.gov.uk/pubns/indg345.pdf

Health and safety training is critical to equipping employees with the skills, knowledge and confidence that they need to succeed in their roles.

The benefits of an employee holding a recognized health and safety qualification include:

- Gaining the skills and knowledge to perform their duties in a safe manner.
- To ensure they are aware of their legal duties as employees under the Health and Safety at Work etc. Act 1974 and Management of Health and Safety at Work Regulations 1999.

- Learn transferable skills which they can use in their workplace:
  - writing reports in areas other than health and safety;
  - learning valuable management skills such as effective communication.
- Becoming more confident in their role.
- The ability to become a member of a professional body.

MEMBERSHIP OF PROFESSIONAL BODIES

Many people who work in health and safety find it beneficial to become a member of a professional body.

The Institution of Occupational Safety and Health (IOSH) was founded in 1945 and is the only chartered body for health and safety practitioners, having more than 40,000 members. It is the world's largest health and safety professional membership organization.

By holding one of the following Level 3 qualifications you can apply to become an associate or technician member.

In the UK these Level 3 certificate qualifications are as follows.

British Safety Council

- Level 3 Certificate in Occupational Safety and Health.
- International Certificate in Occupational Safety and Health.

National Examination Board of Occupational Safety and Health

- Diploma Part 1 (this course is no longer available).
- National General Certificate in Occupational Health and Safety#.
- International General Certificate in Occupational Health and Safety.
- National Certificate in Construction Health and Safety#.
- International Certificate in Construction Health and Safety.
- Fire Safety and Risk Management Certificate#.
- International Certificate in Fire Safety and Risk Management.

Note:# These qualifications are accredited and credit rated on the Scottish Credit and Qualifications Framework (SCQF) at Level 6 which is comparable to NQF/QCF Level 3 in England, Wales and Northern Ireland.

It is possible to gain membership by taking an approved university health and safety qualification or an approved NVQ 3 in Health and Safety.

See http://www.iosh.co.uk for more details.

The International Institute of Risk and Safety Management (IIRSM) was established in 1975 as a professional body for health and safety practitioners. It was created to advance professional standards in accident prevention and occupational health throughout the world. It is a professional membership body that provides recognition, information, support and enhancement for health and safety professionals and specialist members related to the health and safety field.

Membership of IIRSM for associate level may be gained with relevant experience and the following qualifications:

- British Safety Council Certificate in Occupational Safety and Health.
- NEBOSH National and/or International General Certificates in Occupational Health and Safety.
- NEBOSH National Certificate in Construction Health and Safety.
- NVQ Level 3 in Occupational Health and Safety.
- Or an equivalent qualification accepted by the Membership Committee.

See http://www.iirsm.org for more details.

AWARDING BODIES

Details of health and safety courses may be found on the awarding body websites. The most common include the following:

BSC	British Safety Council	http://www.britsafe.org
CIEH	Chartered Institute of Environmental Health	http://www.cieh.org
HABC	Highfield Awarding Body of Compliance	http://www.highfieldabc.com
NEBOSH	National Examination Body of Occupational Safety and Health	http://www.nebosh.org.uk
RSPH	Royal Society for Public Health	http://www.rsph.org.uk

Note: IOSH offers a range of courses that are non-accredited, such as Managing Safely. Details may be found at http://www.iosh.co.uk.

It can be somewhat confusing to work out which qualification would be the most suitable; for example, CIEH, HABC and RSPH Level 3 is similar to NEBOSH Level 2 Award – i.e. all three comprise about 18 hours of study (three days) and have a Multiple Choice Question (MCQ) assessment; but only NEBOSH Level 2 Award has a practical assessment.

NEBOSH General Certificate is the most common health and safety qualification for supervisors and managers. Around 500 Course Providers, in more than 100 countries, now offer NEBOSH courses, with the number of NEBOSH National General Certificate successful students exceeding 150,000.

QUALIFICATION TYPE AND LEVEL

In the UK qualifications are generally set as award, certificate or diploma which indicates the length of the course. One credit represents ten notional hours of learning, showing how much time the average student would take to complete the unit or qualification:

- Awards are 1 to 12 credits (10 to 120 hours' learning).
- Certificates are 13 to 36 credits (130 to 360 hours' learning).
- Diplomas are 37 credits (370 or more hours' learning).

Qualifications and Credit Framework/National Qualifications Framework (QCF/NQF) for England, Wales and Northern Ireland health and safety qualifications are from Level 1 to Level 6.

Source: http://www.direct.gov.uk
http://www.ofqual.gov.uk

NEBOSH qualifications follow the Scottish framework:

> NEBOSH Health and Safety at Work Award has been accredited and credit rated by the Scottish Qualifications Authority (SQA). It sits in the Scottish Credit and Qualifications Framework (SCQF) at SCQF Level 5 with 3 SCQF credit points. SCQF Level 5

is comparable to NQF/QCF Level 2 in England, Wales and Northern Ireland.

Source: http://www.nebosh.org.uk/qualifications/certificate/default.asp?cref=26&ct=2

NEBOSH National General Certificate has been accredited and credit rated by the Scottish Qualifications Authority (SQA). It sits in the Scottish Credit and Qualifications Framework (SCQF) at SCQF Level 6 with 15 SCQF credit points.

SCQF Level 6 is comparable to NQF/QCF Level 3 in England, Wales and Northern Ireland.

Source: http://www.nebosh.org.uk/qualifications/certificate/default.asp?cref=26&ct=2

For more information on different qualification levels see: www.ofqual.gov.uk/files/2009-03-qualifications-can-cross-boundaries.pdf.

Since 2001, mainstream Scottish qualifications have been brought into a single unifying framework known as the Scottish Credit and Qualifications Framework (SCQF). SCQF uses two measures to describe qualifications and learning programmes within the framework:

- The level of the outcomes of learning.
- The volume of outcomes, described in terms of the number of credits.

Source: http://www.sqa.org.uk
http://www.ofqual.gov.uk

BSC qualifications also follow the (QCF/NQF) for England, Wales and Northern Ireland health and safety qualifications are from Level 1 to Level 3 at award and certificate level.

Source: https://www.britsafe.org

CIEH Awards are at levels 1–4:

Level 1	Aimed at those who are new to the workplace and who require a simple training course in a subject to keep them, and those around them, safe when starting work or as part of an induction programme; for example, school leavers taking their first job or those with special needs.
Level 2	Safety training for people in the workplace who participate in activities related to the subject area. The qualification equates to the 'foundation' or 'basic' qualification; for example, shop-floor employees and front-line staff employees in production or services.
Level 3	Aimed at those in the workplace who require safety training courses in activities related to the subject area and who also have a related supervisory role. The qualification equates to the intermediate qualification; for example, team leaders, supervisors and junior managers.
Level 4	Aimed at those in the workplace who participate in activities related to the subject area and who also have a related management and/or training role. The qualification equates to the advanced qualification; for example, senior supervisors, departmental managers and operations managers with responsibility for multiple sites.

Source: http://www.cieh.org/training/qualifications.html

HABC Awards are at levels 1–4:

Level 1	Ensures that candidates are equipped with a basic knowledge of health and safety in their workplace, including the duties under UK health and safety legislation.
Level 2	Provides candidates with knowledge of practices that are essential in the workplace. This includes the hazards and risks presented in the workplace along with the relevant legislation, and ensures awareness of cost-effective, practical control methods.
Level 3	Equips staff to manage the effects of accidents and incidents. It also focuses on the importance of the health and safety culture of a business, which is an invaluable resource to organizations.
Level 4	This qualification is aimed at supervisors and managers with direct responsibility for health and safety management. This qualification focuses on the legal and moral obligations supervisors and managers have to ensure health and safety within a business. It also covers obligations to employees, contractors, visitors and suppliers.

Source: http://www.highfieldabc.com/HealthAndSafety/Default.aspx

Each of the above awarding bodies assesses its course with Multiple Choice Questions (MCQ) and/or written exams. Some also include practical assessments.

INVESTING IN YOUR FUTURE

Health and safety Professional Certificate Qualifications (PCQs) are an investment for your future. Award qualifications do not normally lead to professional membership. Diploma qualifications will lead to a higher membership within the professional body. For example, the NEBOSH General Certificate, with relevant experience, will achieve Technician of IOSH (TechIOSH) status, whereas the NEBOSH Diploma will achieve Graduate of IOSH (GradIOSH) status, which can lead to Chartered Member of IOSH (CMIOSH).

The cost of PCQs will vary depending upon the type of study; for example, distance learning or classroom taught courses. It can also vary from one training provider to another.

'NEBOSH is a name that is recognised throughout the world now for our health, safety and environmental vocational qualifications' (Teresa Budworth, NEBOSH Chief Executive).

CASE STUDY

Richard was a storeman at a local warehouse distribution centre. He enrolled on the NEBOSH National General Certificate course (costing him £1,100) at a local college, one evening a week for 15 weeks, without telling his line manager. After completing the course he presented a copy of his certificate to his line manager and asked if he could get involved with health and safety at the company. Within three months he was promoted to warehouse assistant manager and became head of the company's health and safety committee group.

KEEPING UP TO DATE/CPD AND MOVING FORWARD

Employers are looking for employees who can manage themselves and others.

Initial Professional Certificate Qualifications in health and safety help to build interpersonal skills, people skills and project skills. Once the PCQ has been achieved, it is important to maintain a level of professionalism within health and safety, continually keeping up to date with new initiatives, changes in legislation, etc. It is no longer possible merely to gain a qualification and stop there. A career in health and safety will involve maintaining and updating one's skills and knowledge. This process is known as Continual Profession Development (CPD), which is part of the life of a health and safety manager.

Professional bodies, as previously mentioned, require professional members to record their CPD, to ensure that they are keeping up to date with the ever-increasing changes and demands in health and safety.

One form of CPD could be studying another course. In fact many students study for a range of health and safety qualifications. For example, NEBOSH General, Construction and Fire Certificates all have a common management element known as NGC1. This enables students to combine units across NEBOSH Certificate qualifications. Students do not need to resit any units they have successfully achieved provided that these units were gained within a five-year period for each qualification.

This unit-based structure allows students to:

- Sit all examinations and the practical for the qualification at one sitting.
- Spread their studies over a longer period.
- Simply choose to study a single unit, for example, for CPD purposes.
- Use their previous achievements to attain additional certificate-level NEBOSH qualifications.

See http://www.nebosh.org.uk for more details.

KEY NOTE

Many students take the General, Fire and Construction Certificates, along with the Environmental Certificate, to increase their chances of moving into a career in health and safety.

Undertaking a health and safety course is not the only form of CPD available to health and safety managers. Reading journals, such as SHP online (see http://www.shponline.co.uk), attending seminars or workshops, or reading various free guidance documents published by HSE are all part of CPD.

BEFORE THE COURSE

There are a number of factors that need to be taken into account when identifying which health and safety course you should undertake. These include:

- Previous experience of health and safety.
- Qualifications already held.
- Job role you are going into/currently working in.
- Cost of the course.
- Time needed to study for the course/impact on your work–life balance.
- Recognition of the qualification in the industry you are working in.
- Level of membership the qualification will give you with professional bodies such as IOSH (Institution of Occupational Safety and Health) and IIRSM (International Institute of Risk and Safety Management).

Once you have selected the appropriate course (or your employer may have selected the course for you), it is well worth spending some time trying to find out a little about the course and the course provider. You can do this by looking up the course provider's website and awarding body's websites.

Once you are booked on your course, your course provider will be able to supply you with information regarding the venue, duration of the course, if refreshments are provided, details about the exam, etc. If you have any specific needs it is worth informing the course provider as soon as possible, even if this is before the course. You may even want to ask about the trainer; for example, their experience and qualifications.

CASE STUDY

Martin was booked on a two-week health and safety course paid for by his employer, but he found out during the course that he would not be able to sit the exam. The exam was to be held one week after the course had finished – Martin and his employer had not been made aware of this fact prior to the course beginning and Martin was unable to be released from work to sit the exam. As a result, Martin had to sit the exam some months later.

PRIOR READING

Students may benefit from carrying out some pre-course reading, especially if they do not have a background in health and safety.

General health and safety:

- Successful health and safety management
  http://www.hse.gov.uk/pubns/priced/hsg65.pdf
- Reducing error and influencing behaviour
  http://www.hse.gov.uk/pubns/priced/hsg48.pdf
- Essentials of health and safety at work
  http://www.hse.gov.uk/pubns/priced/essentials.pdf

Fire safety:

- A short guide to making your premises safe from fire
  http://www.communities.gov.uk/documents/fire/pdf/144647.pdf

Construction safety:

- Health and safety in construction
  http://www.hse.gov.uk/pubns/priced/hsg150.pdf

Well-being:

- Working for a healthier tomorrow
  http://www.dwp.gov.uk/docs/hwwb-working-for-a-healthier-tomorrow.pdf

Some courses may recommend or even require prior course reading. If this is the case the course provider should provide you with a reading list prior to the course starting.

Other useful documents for your course include the syllabus, past questions and examiners' feedback. These may be found on the awarding body's websites.

SPECIFIC NEEDS OF STUDENTS

Dyslexia and specific learning difficulties in adults

Dyslexia is a hidden disability thought to affect around 10 per cent of the population, 4 per cent severely. It is the most common of the specific learning difficulties, namely a family of related conditions with considerable overlap or co-occurrence. Together, these are believed to affect around 15 per cent of the population to a lesser or greater extent.

Specific Learning Difficulties (SpLDs) affect the way information is learned and processed. They are neurological (rather than psychological), usually hereditary and occur independently of intelligence.

SpLDs also include:

- Dyspraxia/Development Coordination Disorder.
- Dyscalculia.
- Attention Deficit Disorder.

Dyslexia

> Dyslexia is a specific learning difficulty that mainly affects the development of literacy and language related skills. It is likely to be present at birth and to be life-long in its effects. It is characterized by difficulties with phonological processing, rapid naming, working memory, processing speed, and the automatic development of skills that may not match up to an individual's other cognitive abilities. It tends to be resistant to conventional teaching methods, but its effect can be mitigated by appropriately specific intervention, including the application of information technology and supportive counseling.
>
> Source: British Dyslexia Association

> Ten per cent of the British population are thought to be dyslexic; 4 per cent severely so. Dyslexia is identified as a disability as defined in the Equality Act 2010.
>
> Source: http://www.bdadyslexia.org.uk

The British Dyslexia Association (BDA) campaigns for a dyslexia-friendly society where barriers to dyslexic people do not exist. The BDA works to ensure that all people with dyslexia fulfil their potential.

There are hundreds of dyslexic students who have taken health and safety awards and professional certificate qualifications.

CASE STUDY

Jamie, who is diagnosed as dyslexic, was given extra time in his exam. After speaking to his course provider he gave them a copy of his Dyslexia Statement. Over time he successfully passed the NEBOSH General, Construction, Environmental and Fire Certificates.

Doctor Ian Smythe and John Everett have developed the following checklist.

A checklist for dyslexic adults will not provide enough information for a diagnostic assessment, but it can be very useful in promoting better self-understanding and a pointer towards future assessment needs.

Below are the questions that were found to be more predictive of dyslexia (as measured by prior diagnosis). In order to provide the most informative checklist, scores for each answer indicate the relative importance of that question. Alongside each line you can keep a tally of your score and at the end find a total.

If you think you may be dyslexic you should contact the British Dyslexia Association for further advice.

For each question, circle the number in the box which is closest to your response.

Adult dyslexia checklist

		Rarely	*Occasionally*	*Often*	*Most of the time*	*Total*
1	Do you confuse visually similar words such as cat and cot?	3	6	9	12	
2	Do you lose your place or miss out lines when reading?	2	4	6	8	
3	Do you confuse the names of objects, for example, table for chair?	1	2	4	4	
4	Do you have trouble telling left from right?	1	2	4	4	
5	Is map reading or finding your way to a strange place confusing?	1	2	4	4	
6	Do you reread paragraphs to understand them?	1	2	4	4	
7	Do you get confused when given several instructions at once?	1	2	4	4	
8	Do you make mistakes when taking down telephone messages?	1	2	4	4	
9	Do you find it difficult to find the right word to say?	1	2	4	4	
10	How often do you think of creative solutions to problems?	1	2	4	4	
		Easy	*Challenging*	*Difficult*	*Very difficult*	*Total*
11	How easy do you find it to sound out words such as e-le-phant?	3	6	9	12	
12	When writing, do you find it difficult to organize thoughts on paper?	2	4	6	8	
13	Did you learn your multiplication tables easily?	2	4	6	8	
14	How easy do you find it to recite the alphabet?	1	2	3	4	
15	How hard do you find it to read aloud?	1	2	3	4	

Results from the adults' test: what it all means

The research and development of the checklist has provided a valuable insight into the diversity of difficulties, and is a clear reminder that every individual is different and should be treated and assessed as such. However, it is also interesting to note that a number of questions, the answers to which are said to be characteristics of dyslexic adults, are commonly found in the answers of non-dyslexics.

It is important to remember that this test does not constitute an assessment of one's difficulties. It is just an indication of some of the areas in which you or the person you are assessing may have difficulties. However, this questionnaire may provide a better awareness of the nature of an individual's difficulties and may indicate that further professional assessment would be helpful.

While we do stress that this is not a diagnostic tool, research suggests the following.

Score under 45: probably non-dyslexic

Research results: no individual who was diagnosed as dyslexic through a full assessment was found to have scored under 45 and therefore it is unlikely that if you score under 45 you will be dyslexic.

Score 45 to 60: showing signs consistent with mild dyslexia

Research results: most of those in this category showed signs of being at least moderately dyslexic. However, a number of persons not previously diagnosed as dyslexic (though they could just be unrecognized and undiagnosed) fell into this category.

Score greater than 60: showing signs consistent with moderate or severe dyslexia

Research results: all those who recorded scores of more than 60 were diagnosed as moderately or severely dyslexic. Therefore we would suggest that a score greater than 60 suggests moderate or severe dyslexia. Please note that this should not be regarded as an assessment of one's difficulties. However, if you feel that a dyslexia-type problem may exist, further advice should be sought.

I retook this checklist when writing this book and scored 65!

Course providers should be able to provide you with support. If you have been diagnosed with dyslexia you will need to provide them with a copy of your statement.

ENGLISH AS A SECOND LANGUAGE

The four most common spoken languages (in order) are:

1 Mandarin Chinese.
2 Spanish.
3 English.
4 Arabic.

Many of the awarding bodies will now provide examinations in non-English languages. If English is not your first language you will need to check with your course provider to see if they can arrange for you to sit the exam in another language.

If English is not your first language you may need to take an initial English test prior to the course beginning. This would take the form of IELTS, which is the International English Language Testing System, the world's proven English-language test. More details may be found at: http://www.ielts.org.

Students who take award-level qualifications (Levels 1–3) would benefit from having an IELTS score of 5.0 or higher.

Students who take award Level 4 or certificate qualifications would benefit from having an IELTS score of 6.0 or higher.

AT THE START OF THE COURSE

It is important that you bring with you pens, pencils, highlighter and something to write in (e.g. a spiral A4 notebook). Some training providers will give you these items on the first day of the course; however, if they don't, it could be embarrassing if you have to 'borrow' some items.

If possible, arrive early at the course venue. This will give you time to orientate yourself with the facilities.

At the start of the course it is likely that your trainer will introduce him- or herself and make time for members of the group to introduce themselves to each other. This will be an important time to identify the other students, where they work and the experience they have in health and safety. For example, if one of your fellow students is a firefighter, they may be able to provide additional information during the course, or clarify something that was covered in class.

Your trainer(s) should have the necessary qualifications, skills and experience to deliver the course. There is nothing wrong in asking the trainer about his or her qualifications, skills and experience that enable him or her to deliver the course.

CASE STUDY

As a trainer I have had the privilege of delivering health and safety courses in the UK, Africa and America.

One of my most memorable training experiences was in Nigeria. It was the beginning of my first-ever NEBOSH International General Certificate course. At the start of the course I gave the normal introduction, explained how the course was to be structured, when the exams would be, etc.

At this point one of my students raised his hand and asked (in so many words): 'What qualifies YOU to deliver this course to us?'

I was amazed at his bluntness, and yet more amazed that I had never been asked this question before. I had already explained that this was my first international course. After a moment taken to compose myself I started by congratulating him on the best question I had ever been asked. I was able to tell him that I had delivered over a dozen NEBOSH Certificate courses in the UK and that the syllabus was very similar to the national course. I also explained that I hold a Certificate in Education and BA(Hons) in Education and Training. I also explained that I was a NEBOSH examiner for both national and international papers.

UNDERSTANDING MEMORY

The brain has been designed to receive and process information. Our knowledge is not localized in any one part of the brain. Connections are made between the neurons as you receive the new health and safety information.

Millions of connections take place in the brain as you study. All of the information gained during your study will reinforce the connections (pathways) or create new ones. This makes it possible to think about any aspect of health and safety and follow a train of thought (pathway) to other topics you have studied. For example, if you have been studying personal protective equipment (PPE) you might think about the types of work that would require specific items. Therefore you might start thinking about the PPE used in your own workplace.

EXAMPLES

While listening to their trainer discussing fire safety, a class of students will also be thinking about and making connections with previously held memories; for example:

- Adam remembers setting a chip pan alight and struggles to focus on what is being taught.
- Barry thinks about his new office and that he does not yet have any fire extinguishers.
- Catherine, who works in a care home, focuses her thoughts on how residents would get out of the building in an emergency.

Each of the three students receives the same information. However the connections that are made in the brain between the neurons will be different depending on each student's previous experience.

Later, when the students are thinking about the topic of fire during revision, they are likely to remember their thoughts at the time of being taught the theory. For example, Barry may picture his office and think about which fire extinguishers are needed.

CASE STUDY

After class Joshua would read through his company's health and safety policy. He would make notes on the policy relating to what he had previously been taught in class. This allowed him to link the course information to his own company.

The successful study of health and safety depends upon how much information you can take in, giving the brain time to make all the millions of connections. Therefore the more information/experiences you receive about a particular topic the more connections will be made/ reinforced, so that you will be able to recall what you have learned at a later time.

If the information received can be linked together you are more likely to be able to remember and recall it when necessary. This can be illustrated by the following two activities.

ACTIVITY 1

Study the following list for two minutes and then try to write down the items without looking back at the list:

Alarm	Foam
Burning	Fuel
Carbon dioxide	Hose
Conduction	Ignition
Convection	Oxygen
Door	Powder
Endothermic	Radiation
Exothermic	Water
Flashpoint	Wet chemical

Which words did you remember:

- Words at the beginning of the list or at the end?
- Words that were already familiar to you?
- Words that sounded strange?
- Words that you could picture, for example, a door?

ACTIVITY 2

By putting the list into some sort of order (i.e. linking words together), you are more likely to remember the items on it. Study the following list for a further two minutes and then try to write down the items without looking back at the list:

Water
Powder
Carbon dioxide
Wet chemical
Foam

Fuel
Ignition
Oxygen

Conduction
Convection
Radiation
Burning

Endothermic
Exothermic
Flashpoint

Door
Alarm
Hose

By linking words together you are more likely to remember them.

COMMUNICATION

Communication is a two-way process between a transmitter and receiver. In verbal communication the transmitter and receiver would be the trainer and the student, or vice versa. In written communication the transmitter would be the course material and the receiver the student. Communication may break down if the receiver is unable to understand what is being conveyed.

Within the context of the learning environment there will be barriers to communication and therefore barriers to learning. These include:

- Hearing what we expect to hear.
- Ignoring information which conflicts with what we already believe/know.
- Perceptions about the communicator.
- Influence of the learners.

- Words having more than one meaning.
- Emotional context.
- Poor learning environment (e.g. too cold/hot, noise, etc.).
- Different accents.
- Language difficulties.

Communication is based not just on the words we speak but on how they are expressed: 'It is not what we say, it is the way that we say it!'

Communication is:

- 7 per cent – what we say.
- 38 per cent – how we say it.
- 55 per cent – body language.

It is important that you are able to communicate with your trainer. If you are unsure of something you hear, see or read you will need to seek clarification. If you have any questions you will need to ask your trainer as soon as possible – other students may need clarification too.

GROUP DYNAMICS

Your fellow students will be a great source of support, and likewise you too can support them. As part of a normal teaching classroom experience you should expect to work together in small groups or in pairs, and you will be asked to feed back your findings to the whole group (i.e. an andragogy style of teaching; namely learning strategies focused on adults, often interpreted as the process of engaging adult learners within the structure of the learning experience).

In 1965 Bruce Tuckman unveiled what would become the standard way of describing group dynamics. Tuckman describes four phases of group development: forming, storming, norming and performing (later adding a fifth stage: adjourning).

1 **Forming** The initial stage where learners may be anxious and need to know boundaries. Learners may not feel safe.
2 **Storming** Conflict between learners and the trainer may arise.
3 **Norming** This is when cohesion develops and the learners start to cooperate.

4 **Performing** Learners now feel safe and are able to work together with their fellow learners.

5 **Adjourning** Learners often leave a group with the desire to meet again. This tends to happen when a course is delivered over an extended period of time.

It makes for a much better learning experience if students are focused in their study, all wanting to succeed.

If you have any concerns about the teaching/fellow students, etc., you must address them with the trainer as soon as possible so that the trainer is able to remedy matters quickly. It is pointless, both for you and your group, to wait until the end of the course before registering your concerns. It will be too late to rectify matters, and addressing them early will make your learning experience more enjoyable and beneficial.

TIME MANAGEMENT

A successful student is organized and focused. It will be of benefit if you can map out study and revision time at the start of your course. The following pages give some examples of the types of study plan that may benefit you if you are studying for a professional certificate qualification.

Sometimes this will not be needed; for example, if you are starting a one-week course with the exam taking place on the Friday afternoon, your revision time will be Monday to Thursday evenings, which your trainer should set for you.

Some courses may run for one day per week. This is when a study diary or timetable will be beneficial. These are also crucial when studying any course distance learning.

Developing your own timetable can be beneficial, as this will give you ownership. Likewise, it is important that you focus your revision on the areas you need (i.e. areas where your knowledge is weakest).

Remember: you are making an investment in yourself, so you may need to cancel any plans you have during the course; for example, missing that football match shown on television.

Example 1: Two-week course

The following is an example of a two-week revision timetable.

Day	*Morning*	*Afternoon*	*Evening*
Monday	Course		Homework from trainer
Tuesday	Course		Homework from trainer
Wednesday	Course		Homework from trainer
Thursday	Course		Homework from trainer
Friday	Course		Homework from trainer
Saturday	Mock exam 1	TIME OUT	Review mock
Sunday	TIME OUT	Revise week 1	Revise week 1
Monday	Course		Homework from trainer
Tuesday	Course		Homework from trainer
Wednesday	Course		Homework from trainer
Thursday	Course		Homework from trainer
Friday	Course		Homework from trainer
Saturday	Mock exam 2	TIME OUT	Review mock
Sunday	TIME OUT	Revise week 1 and 2	Revise week 2
Monday	Exam	Exam	Go to the pictures!

Example 2: Day-release course

Week number	*Material covered/unit*	*Revision time planned*	*Made revision notes*	*Reviewed notes*	*Practice questions*	*Confident*
1						1 2 3 4 5
2						
3						
4						
5						
6						
7						

8						
9						
10						
11						
12						
13						
14						
15						

Mock 1 DATE ________ Mock 2 DATE ________ Exam DATE ________

Score yourself on how confident you are; keep going back until you score yourself 5.

Example 3: Distance learning

Week number	*Material covered/ unit*	*Revision time planned*	*Made revision notes*	*Reviewed notes*	*Practice questions*	*Confident*
1						1 2 3 4 5
2						
3						
4						
5						
6						
7						
8						
9						
10*						

Mock 1 DATE ________ Mock 2 DATE ________ Exam DATE ________

Score yourself on how confident you are; keep going back until you score yourself 5.

Note: * Number of weeks will depend upon:

- Number of units/elements to cover.
- Number of hours recommended.
- Date of exam.

Example 4: Focused study plan

Learning outcome	*Receive information*	*Remember information*	*Recall information*	*Confident*
				1 2 3 4 5

Mock 1 DATE ________ Mock 2 DATE ________ Exam DATE ________

Score yourself on how confident you are; keep going back until you score yourself 5.

Number of learning outcomes will be identified on the course syllabus.

ORGANIZE YOUR WORKING ENVIRONMENT

You will have little choice about the classroom, but remember: you must tell your trainer if there is an issue (e.g. is the classroom too hot or too cold?). If you need special access (e.g. for a wheelchair), you should inform your course provider prior to the course starting if at all possible.

When you are studying outside of the classroom, perhaps at home or even in a hotel (if you are away on a residential course), you will need to make sure that your study area is suitable for your needs. It is important that you have a comfortable chair, enough space to work, access to a clock/watch and a drink (e.g. a bottle of water or a coffee). You may like to have background noise – music or the radio – or you may prefer to work in a quiet environment. Make sure the room is at a comfortable temperature, has good lighting and adequate ventilation, as these factors can have either a positive or a negative impact on your study time.

It is easy to get distracted by phone calls, someone at the door, emails, etc. One way to reduce the amount of distractions is to have your mobile on silent, switch on your answerphone on your land-line and switch off your email program.

DISTANCE LEARNING

Distance learning/e-learning courses will involve you studying on your own, setting your own timetable, etc.

A large course book, e-book or portable document format (PDF) file for a health and safety course can be quite daunting. While it is important to understand the details of the course, it may be of benefit to have an overall understanding of the big picture first so that you can see how they all link together. It is well worth having an overview of the whole course before you start; this can be accomplished by reviewing, first, the learning outcomes for each section and then skimming through the text to see the areas that will be covered on your course.

Health and safety covers a wide range of topics, including:

- Legal – civil and criminal/common and statute.
- Management systems/policies/risk assessments.
- Human factors, including culture.
- Emergency procedures/first aid and accident investigation.
- Auditing and measuring health and safety performance.
- Chemical/biological.
- Construction.
- Electricity.
- Ergonomics.
- Fire.
- Noise/vibration.
- Radiation/temperature.
- Stress/violence/substance misuse at work.
- Transport.
- Work equipment.
- Working at a height.
- Workplace safety, health, welfare and environmental factors.

It is important to break down the topics into manageable chunks of information that you can understand. Rather than tackling the entire course syllabus in one go, it is easier to address one section at a time. The best way to do this is to use the learning outcomes as natural small sections.

CASE STUDY

Steve copied out each of the learning outcomes from the syllabus on to the tops of fresh A4 pages in his spiral notebook, and then made notes from the e-learning program supplied by his course provider.

RECEIVE

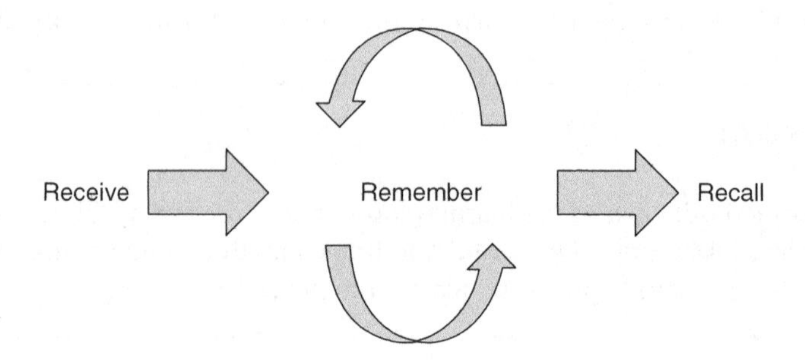

All study/learning, whether a formal course or simply learning a new hobby, can be split into three areas:

1. Receive the information (new and previously known).
2. Remember the information (through retaining, reviewing, rehearsing and retrieving).
3. Recall the information (for an exam or completion of a task).

The following three sections will outline the processes involved within each of the three Rs, provide positive examples of applying each R, offer key notes, and furnish you with the essential study skills for taking a health and safety course.

These three areas and the principles described may be applied to other study areas. They have been developed over the past ten years or so during my own study and teaching.

During your health and safety course you will receive information in several ways, including:

- Listening to your trainer and fellow students.
- Reading course book(s), syllabus, past exam questions, etc.
- Seeing videos, signs, posters, etc.
- Taking part in activities/practical exercises.

You will be required to recall specific parts of this information for your exam. To do so you will need to develop strategies to remember the necessary information.

It is important to filter in the information you need and to filter out the unwanted information. For example, during the course you may be given a large amount of detailed information (e.g. the specific content of different sizes of first aid kits). While it is important that you know the general content and more specifically what should not be included in the kit (drugs and medication), specific knowledge of the number of plasters, bandages, etc. is not normally required.

The syllabi and past/exemplar questions would indicate the level of detail that is needed. Your trainer should explain the depth and breadth of knowledge needed.

EXAMPLES

The following are some examples/case studies of how students have actively received the information taught in class:

- Writing down the main points from the course book and PowerPoint while in class.
- Drawing lots of pictures/sketches relating to what has been taught.
- Making notes in and highlighting certain information in the course book relating to what has been taught.

HOW TO MOTIVATE YOURSELF TO STUDY

Regardless of your experience in sitting exams at school or college, it is important to be confident in your own ability. To pass any exam, regardless of its being health and safety, takes commitment from you, the student.

A student who has a positive reason for wanting to learn is more likely to succeed and to retain what he or she has learned.

ACTIVITY

What motivates you to study?

Maslow first introduced his concept of a hierarchy of needs in his 1943 paper 'A Theory of Human Motivation' and in his subsequent book *Motivation and Personality*.

This hierarchy suggests that people are motivated to fulfil basic needs before moving on to other, more advanced needs. Maslow sets out five levels through which a student can progress. Before you can progress to the next stage your needs at the current level must be achieved.

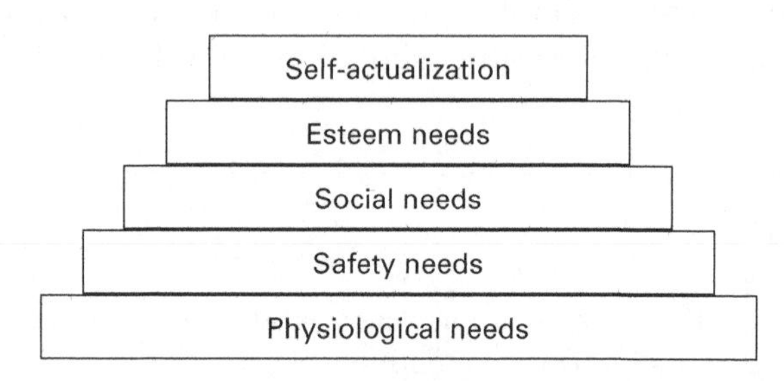

Self-actualization	Provide students with a challenge and the opportunity to reach their full learning potential.
Esteem needs	Recognize achievements in order to make students feel appreciated and valued.
Social needs	Create a sense of community within the classroom.
Safety needs	Provide a safe learning environment and security.
Physiological needs	Provide lunch breaks, rest breaks and an explanation of what is required both during and after the course.

During the course the trainer should work towards meeting the esteem needs of the group by encouraging students throughout the course. Likewise, the trainer should help you to reach your potential; in practice this could be simply to guide you towards taking a higher level course once you have finished the one you are currently taking.

You also have a role to play in these stages; for example:

Self-actualization	Can I do the best I can?
Esteem needs	Can I encourage fellow students as well as myself during the course?
Social needs	Can I create a sense of community within the classroom, so that we can support one another?
Safety needs	Can I make sure I do not cause a danger to the class?
Physiological needs	Can I ensure that I have sufficient food and drink, etc. for the course?

CASE STUDY

Ashley (who is dyslexic) took his NEBOSH Certificate a number of years ago. He passed. This gave him the confidence to take the NEBOSH Diploma, which he also passed.

ACTIVITY

What are your personal goals for the course you are about to undertake?

At the end of my ______________________ course I want to:

Gain the grade of: Pass/Credit/Distinction or Merit.*

Developing the knowledge and skills involved in this course will enable me to:

____
____
____
____
____
____
____
____
____
____
____
____
____

Take my study further by sitting the:

____ certificate.

Note: * some awarding bodies use Pass and Merit while others use Pass/Credit and Distinction.

UNDERSTANDING THE SYLLABUS

Before you start your course it will be helpful to look at the syllabus or syllabus summary. These documents may be found on the awarding bodies' websites, or are available from your course provider.

A syllabus consists of the learning outcomes for each section/unit/element of your course. It may also provide details of how the course is assessed and the time needed to cover each section, both in class and for private study.

SYLLABUS EXAMPLE

The following is an example of what may be included in a syllabus for the element of manual handling within a health and safety course in the UK.

Time: Two hours.

By the end of the session the student should be able to:

- Explain the duties of the employer within the Manual Handling Operations Regulations 1992.
- Identify the types of injury associated with manual handling.
- Identify the correct stages to be used when lifting.
- Outline the principles of carrying out a manual handling risk assessment.

The above example provides you with the details that are to be covered in a two-hour session.

If your assessment is multiple choice, a suitable question would be:

EXAMPLE 1

Which are the stages of a manual handling risk assessment?

A Task, Individual, Load, Environment.
B Training, Individual, Load, Environment.
C Hazard, Risk, Control.
D Task, Individual, Lift, Environment.

(Answer: A)

It would not include questions about other types of risk assessments:

EXAMPLE 2

Which is the first stage of the five steps to risk assessment?

A Record significant findings and implement them.
B Identify the hazards.
C Evaluate the hazards.
D Review and update as necessary.

(Answer: B)

Example 2 should not have been used, as this information is not taught in this part of the syllabus (i.e. there is no learning outcome 'Outline the five stages of a risk assessment' for this mini-syllabus).

If your assessment is a written exam, possible questions may be:

Identify	types of manual handling injuries in the workplace.
List	four areas that are assessed during a manual handling risk assessment.
Define	the term 'manual handling'.
Describe	the duties placed on an employer under the Manual Handling Operations Regulations 1992.
Outline	the correct technique for manual handling.

Questions that relate to manual handling but would not be used in a written exam because they are not within the learning outcomes include:

Give	examples of those who may be involved in assessing an employee who has injured their back at work.
Name	the five regions of the spine.
Explain	the term upper limb disorders (ULDs).
State	the reasons why a manual handling incident should be investigated.
Compare and contrast	sprains and strains.

The reason why the last question '**Compare and contrast** sprains and strains' could not be used as an assessment question is that the learning outcome was to '**Identify** the types of injury associated with manual handling'. While it is true that sprains and strains may result from a manual handling injury, to expect students to be able to compare and contrast this requires more understanding than the knowledge sought by the learning outcome.

The trainer is likely to use a mixture of teaching techniques, including lectures, group work and individual work. The trainer overseeing this two-hour session may elect to include a practical, which would include, based on the learning outcome, the need to identify the correct stages to be used when lifting. This would be a practical exercise involving students moving a small load (e.g. a ream of A4 paper) from the floor on to a table.

A manual handling course may require that the students participate in a specific lift. If this is the case then this would be included within the learning outcomes (e.g. using a safe lifting technique to move an object from the floor on to a table). An alternative would be to carry out a safe practical lift of a specified object.

Having your own copy of the course syllabus will benefit you by being able to:

- Know the areas that are covered and the depth of study needed.
- See how much time should be allocated to each part of the course.
- Make an assessment of the types of questions that may be asked.
- Use the syllabus as a revision aid.

ACTIVITY

As you read the syllabus for the course you are sitting/about to sit, ask yourself: What is likely to be covered in the course/course book?

ACTIVE LEARNING

During the study stage you will need to receive the information – either through attending a course or by studying at home through distance or blended learning.

Passive learning is by simply listening to a lecture or reading a book. You are unlikely to remember much of what you simply hear or read.

Active learning requires participation, including making notes during a lecture or highlighting a book you are reading; following this you could write up your course notes and prepare a summary of the book/chapters you have read. The more you do, the more you will remember.

During the 1960s, Edgar Dale theorized that learners retain more information by what they 'do' as opposed to what is 'heard', 'read' or 'observed'. His research led to the development of the Cone of Experience:

10 per cent of what they read.
20 per cent of what they hear.
30 per cent of what they see.
50 per cent of what they see and hear.
70 per cent of what they say and write.
90 per cent of what they do as they perform a task.

Source: Dale, *Audiovisual Methods in Teaching* (1969)

This was developed further by Mel Silberman, who stated in his book *Active Learning: 101 Strategies to Teach Any Subject*:

What I **hear,** I forget.
What I hear and **see,** I remember a little.
What I hear, see, and **ask questions about** or **discuss** with someone else, I begin to understand.
What I hear, see, discuss, and **do,** I acquire knowledge and skill.
What I **teach** to another, I master.

KEY NOTE

Some students struggle to listen to the trainer at the same time as reading the PowerPoint and/or course notes/book while making notes.

CASE STUDY

While sitting a two-week NEBOSH health and safety course, Paul would read the material to be covered the following day from the course book the night before. This way, when he listened to his trainer he already had some understanding. He made brief notes in class and wrote them up before reading the following day's material.

ACTIVE LISTENING

Have you ever listened to someone talking and then realized you have not 'heard' a word they are saying? Has your mind ever wandered?

In a health and safety class you will be expected to listen, participate and take notes.

According to psychologist Dr Rob Yeung, the average person's attention span is approximately 20 minutes.

There are many factors that will affect your attention span in the classroom. These include:

- Currently under stress (from work or home life).
- Tired – due to lack of sleep.
- Been unwell.
- Not interested in the course.
- Having a poor trainer.
- Feel that the course is too hard or too easy.

Your attention span is different depending upon what you are doing – if you have a keen interest in something your attention span is likely to be longer than 20 minutes.

As a health and safety student you will need to ensure that you try and keep your attention span as long as possible. This may be achieved by having enough sleep before and during the course, looking after your health, maintaining your interest by keeping focused on the course material, and actively thinking of ways to apply the information received to your workplace/job role.

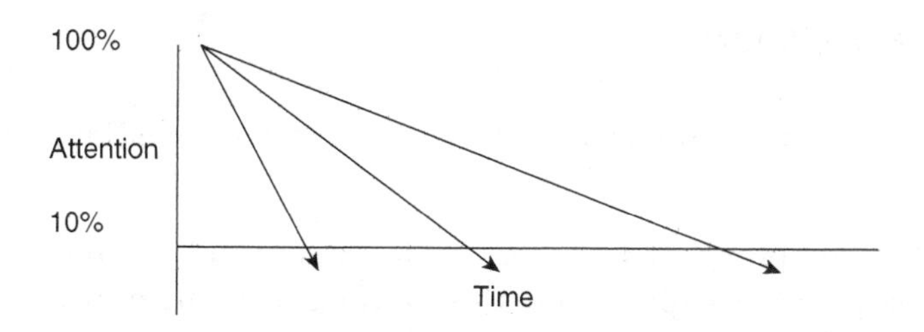

The three arrows show how long a student's attention span may last. It is true that the trainer may be able to increase students' attention span with comments like 'Now, to summarize this session' or 'This comes up in exams often ...'.

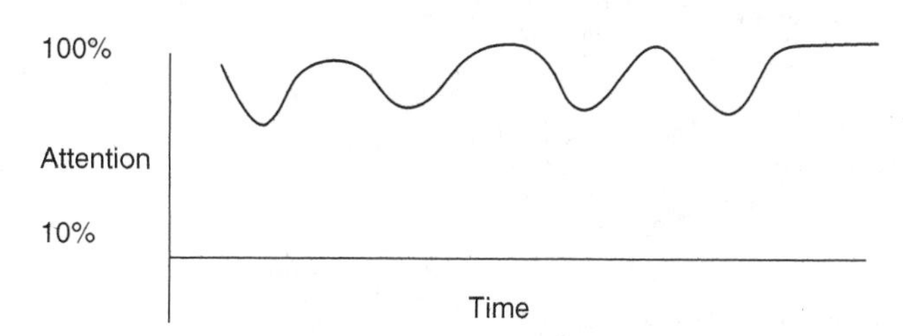

It is impossible to keep one's attention focused 100 per cent of the time. The key to effective listening is to reactivate your attention by proactively doing something. Making notes is one of the most obvious ways you can do this.

If you have to listen to something that is not relevant (e.g. anecdotes), instead of switching off:

- Make a note of a more appropriate example – something that is relevant to you.
- Read over your notes prior to the start of the next section and add more information/underline key points.
- Look back at the learning outcome from the syllabus for the section you are working on and make some extra notes.
- Ask your trainer: 'How does this apply to the syllabus?'

It is important that what you do is related to the section you are covering in class.

ACTIVE NOTE TAKING

There is no right or wrong way to make notes; some are more useful than others.

Some students prefer to write out in linear form, i.e. listing each new item on a piece of paper. Others use a nuclear or pattern note-taking method (e.g. Spider diagram, Concept Map or MindMap ® (Tony Buzan)).

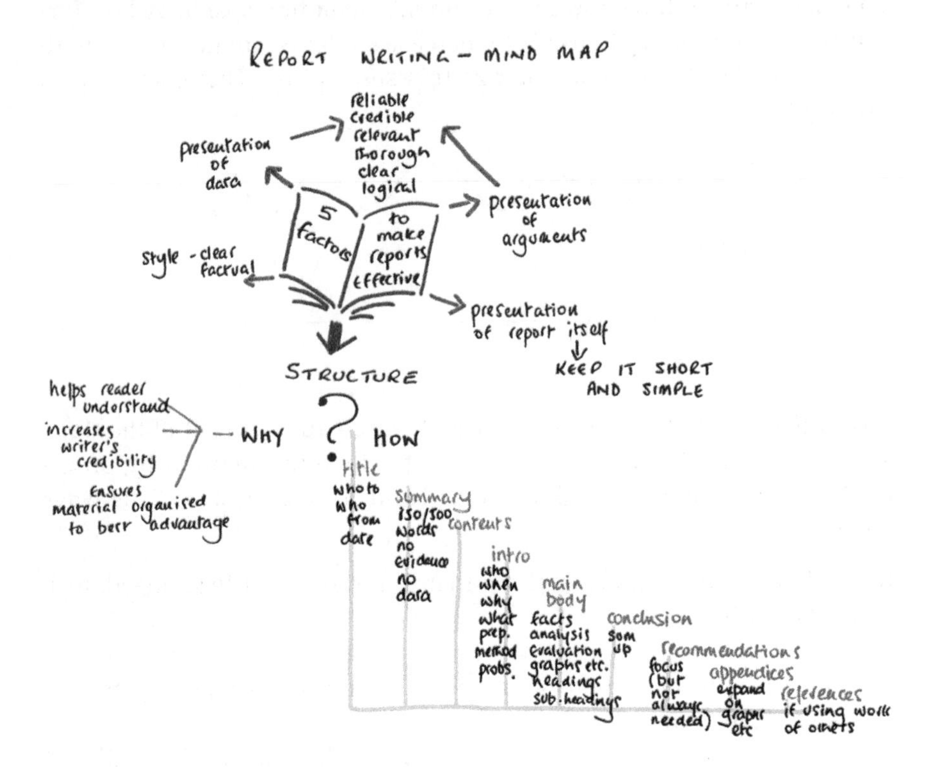

(Source: Hughes and Ferett, 2012)

Simple Spider diagrams can be constructed quickly in class and may be used as a way of receiving information.

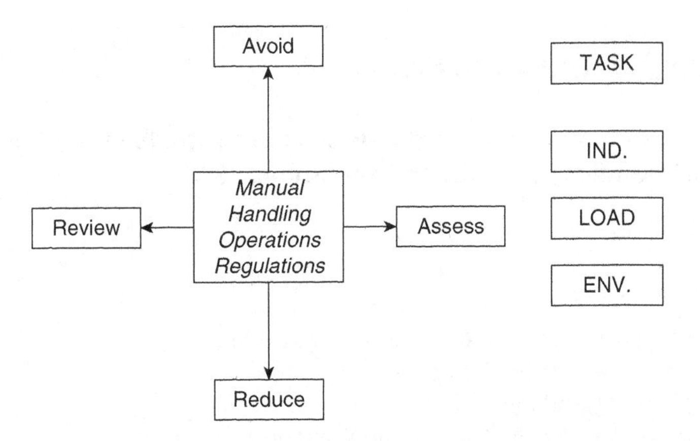

Sometimes it can be difficult to take effective notes in class. The following list provides you with some guidance on making effective notes:

- Be selective when making notes – focus on writing down the main facts.
- Use abbreviations such as PPE for personal protective equipment or shorten words (e.g. emp'er for employer).
- Ignore spelling, grammar and punctuation when note taking.
- When writing out a quote or a piece of text – if you can access it afterwards leave sufficient space and make a note of where to look it up later. (This will help you with the second stage of the three Rs study technique – to Remember).
- Remember: you can ask your trainer to slow down or repeat something.

It is also important to remember that your notes are *your* notes: as long as you can understand what they say then they are more than adequate.

Notes can be made in a linear format: see example below. These could be notes made during a manual-handling two-hour training session.

> **KEY NOTE**
>
> At the start of each new section of the notes the student should write down the learning outcome.

EXAMPLE OF LINEAR FORMAT

Learning outcome: Explain the duties of the employer within the Manual Handling Operations Regulations 1992

1992 No. 2793

Part of six pack:

- L21 Management of Health and Safety at Work.
- L22 Safe Use of Work Equipment.
- L23 Manual Handling.
- L24 Workplace Health, Safety and Welfare.
- L25 Personal Protective Equipment at Work.
- L26 Work with Display Screen Equipment.

READ: indg143

Is the transportation or support of a load by:

- Lifting.
- Putting down.
- Pushing.
- Pulling.
- Carrying.
- Moving.

by hand or by bodily force?

Employer

Employer to minimize the health risks associated with manual handling.

Reg 4:

- Avoid hazardous manual handling operations so far as is reasonably practicable.
- Assess any hazardous manual handling operations that cannot be avoided.
- Reduce the risk of injury so far as is reasonably practicable.
- Review.

AARR = Avoid/Assess/Reduce/Review.

Need to record significant findings?

While at work employees shall make full and proper use of any system of work provided.

Learning outcome: Identify the types of injury associated with manual handling

The spine is made up of 24 bones called VERTEBRAE.

Poor manual handling is the most common cause of work-related injury, and the most common site of injury is the back. Back injuries can result from a single wrong movement, but more commonly result from repeated stress on tired muscles.

- Slipped/prolapsed disc.
- Sprain.
- Strain.
- Dislocation.
- Fracture/break.
- Cuts/abrasions.
- Hernia/ruptures.
- Crushing/impact injuries.

Seven students out of nine in class have bad backs, including the trainer! It is estimated that one-third of the workforce suffer with back problems.

Hazardous work:

- Nurses.
- Postal workers.
- Airport workers.
- Weight-lifters!

Learning outcome: Identify the correct stages to be used when lifting

STRAIGHT BACK IS WRONG (last course was taught this?).

- Think before lifting/handling.
- Keep the load close to the waist.
- Adopt a stable position and get a firm hold.

- At the start of the lift, slight bending of the back, hips and knees is preferable to fully flexing the back (stooping) or fully flexing the hips and knees (squatting).
- Don't flex the back any further while lifting.
- Keep the head up when handling and move smoothly.
- Don't lift or handle more than can be easily managed.
- Put down, then adjust.

Equipment: sack truck/trolley, etc. (training?).

Learning outcome: Outline the principles of carrying out a manual handling risk assessment

Who can teach manual handling?

Read Schedule 1

Factors for which the employer must have regard and questions to consider when making an assessment of manual handling operations:

TILE or LITE

- **T**ask holding away, twisting, stooping, rate of ... etc.
- **L**oad heavy, bulky, etc.
- **E**nvironment space, floors, levels, environment – temperature, ventilation and light
- **I**ndividual unusual strength or height needed, pregnant, medical, information and training
- Other factors PPE, etc.

See MH risk assessment handout.

You will have noticed that additional information was written down during the session. Additional information was also written down. Remember: even though this may not be part of the syllabus it may be helpful for recalling the necessary information.

For example: STRAIGHT BACK IS WRONG! was written down and the correct technique was then copied out. If the student thinks about the reasons why this is wrong, why it is a common error that is taught, he or she is likely to remember the correct technique.

Notes may also be copied out on to index cards so that you can carry them with you.

TILE OR LITE

Task	holding away, twisting, stooping, rate of ... etc.
Load	heavy, bulky, etc.
Environment	space, floors, levels, environment – temperature, ventilation and light
Individual	unusual strength or height needed, pregnant, medical, information and training.

CASE STUDY

When making notes in class, Graham would write actions to be carried out afterwards and circle them – this way he could write down all the necessary information and develop his notes later:

Look up full title of Regulation

from www.

Look up manual handling trainer.

List other examples

Read work policy

Make notes

Remember to ask yourself if you understand what you are writing or what has been taught. There is no such thing as a stupid question! If you are unsure of what the trainer is saying there is nothing wrong in asking questions or even challenging, respectfully, what he or she says. If you are studying a course through distance learning you should be able to phone or email your trainer.

CASE STUDY

Kelly would read a chapter of her course book and then take a break. When she went back to read the next chapter she would first re-read the headings of the previous chapter, and think about what she had read before moving on.

REMEMBER

You have now received the information you need to develop the strategies you already use to ensure you remember the course material.

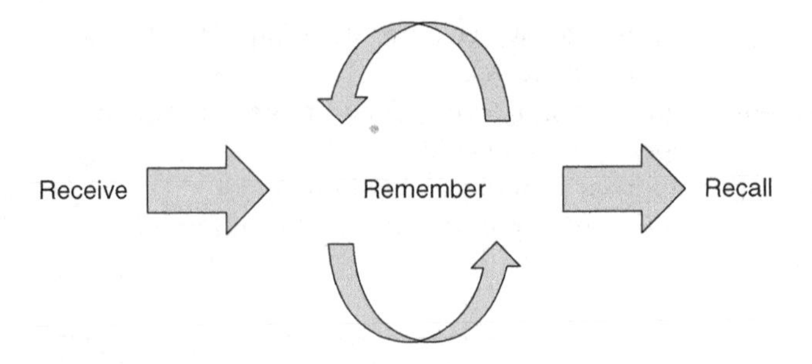

The process of remembering is drawn as a loop because it is a process you will be involved in of continually reviewing, revising, rehearsing and retrieving the information as you begin to remember the material. The information must be moved from your short-term memory to your long-term memory where you can recall it at any given time, including during the examination (see the following section, 'Recall', for more details). The time spent, and even the techniques used, will vary from one student to the next. This section will offer a range of examples that some students have found useful. Some of these techniques will be more effective than others. There are no right or wrong techniques. Every student is different; therefore you need to work out for yourself which are the most effective techniques.

CASE STUDY

Diane recorded notes on to her iPod to listen to in the car as she drove each week to her health and safety course.

CASE STUDY

Jackie would copy out the key notes when she returned home on each night of the course. Just before the exams all she needed to do was to go over her new notes without having to look up her class notes, PowerPoint or course book.

WAYS STUDENTS REMEMBER INFORMATION

While there are different theories identified below of how we study, it is important to understand that throughout your life you have developed your own way of studying; it is unique to you. What works for one person may not work for someone else. For example, when studying, some students find it helpful to have music on in the background, while others need complete silence.

The following are some examples/case studies of how students actively remember the same course material:

- Recorded the book on to an MP3 and played it back numerous times.
- Drew lots of pictures of different aspects of health and safety and labelled them.
- Wrote out key words and phrases on to index cards and kept reading them over and over again.
- Completed as many past exam questions as possible.
- Read through past examiners' reports after completion of questions from those papers.
- Tried to imagine that they were talking to their boss about implementing health and safety issues – and having to persuade him or her.
- Took photographs of as many health and safety items as possible; as they edited them in Photoshop they would be thinking about that particular issue.
- Rewrote their company's health and safety policy, along with risk assessments, during a course (helping them to apply the information taught).
- Downloaded documents from the HSE website about the topics covered and read them in addition to the course material.
- Wrote mini-stories about aspects of the course (see Example 1).
- Reviewed the list of duties of an inspector and imagined they were the inspector going round their place of work (see Example 2a and 2b).
- Read through their notes and made a glossary of words they were not familiar with or which had a specific definition (see Example 3).
- Found an example of what had been discussed in class (see Example 4).
- Looked at each step of the five steps to risk assessment and then applied them to their workplace (see Example 5).
- Looked at each step of the five steps to fire risk assessment and then applied them to their workplace (see Example 6).

- Looked at different safety signs in the workplace (see Example 7).
- Looked up case studies for examples (see Example 8).

If you write up your class notes, or look up certain information as soon as possible, this will help you remember more of the information you were taught. The rewritten notes will also be useful for reading over before the exam.

EXAMPLE 1: MAKING A STORY

A night on the town ... to help you remember the Management of Health and Safety at Work Regulations (starting at Reg 5) sections in order.

The story...	Regs	Sections
Arranging a night out with some mates to survey the nightclubs; you need some assistance in this.	5 6 7	H & S arrangements Health surveillance H & S assistance
There are a few dangerous areas, so you will need to contact a taxi firm; give them the information they need so that they can cooperate and coordinate at their end.	8 9 10 11	Serious and imminent danger Contact with external services Information Cooperation and coordination
Are any of them bringing mates and are they all capable or do they need training? If all are OK, give them their duties.	12 13 14	Host employers and undertakings Capabilities and training Employees' duties
Are any coming for only part of the night or are any of them pregnant? If so, they will need to bring their certificate to notify me.	15 16 17 18	Temporary workers Expectant mothers (risk assessment) Expectant mothers' certification Expectant mothers' notification
Finally, are any under age? If so, they are exempt as they will be a liability.	19 20 21	Young persons Exemptions Provision as to liability

EXAMPLE 2A: LIVING AS A HEALTH AND SAFETY INSPECTOR

Powers of inspectors – imagining yourself in the role will help you to remember the information needed.

ACTIVITY

Ask yourself the questions posed in Example 2b and imagine yourself in this role.

Health and Safety Inspectors under the Health and Safety at Work etc. Act 1974, either for HSE or a local authority, are authorized to enter premises at any reasonable time (or at any time in a dangerous situation), and to:

- Take a constable with them if necessary.
- Take with them another authorized person and necessary equipment.
- Examine and investigate.
- Require premises, or anything in them, to remain undisturbed for the purposes of examination or investigation.
- Take measurements, photographs and recordings.
- Cause an article or substance to be dismantled or subjected to any test.
- Take possession of, or retain anything, for examination or legal proceedings.
- Take samples as long as a comparable sample is left behind.
- Require any person who can give information to answer questions and sign a statement.
- Request information, facilities, records or assistance.
- Do anything else necessary to enable them to carry out their duties.
- Issue an improvement notice.
- Issue a prohibition notice.
- Initiate prosecutions.
- Seize, destroy or render harmless any article or substance which is a source of imminent danger.

EXAMPLE 2B: AN ACTIVE IMAGINATION WOULD GENERATE THE FOLLOWING QUESTIONS

Options	Questions to ask
Take a constable with them if necessary.	Which constable would you take – someone off TV perhaps?
Take with them another authorized person and any necessary equipment.	Who else could I take – which of my friends could help me?
Examine and investigate.	What is in my workplace?
Require premises, or anything in them, to remain undisturbed for purposes of examination or investigation.	How would this affect the company? Would it stop all work?
Take measurements, photographs and recordings.	What could I measure? What would make a good photograph?
Cause an article or substance to be dismantled or subjected to any test.	What could I take apart?
Take possession of, or retain anything, for examination or legal proceedings.	What could be taken?
Take samples as long as a comparable sample is left behind.	What samples should be taken?
Require any person who can give information to answer questions and sign a statement.	What questions would I want to ask and to whom?
Require information, facilities, records or assistance.	Whose office would I want to occupy while I carry out the investigation?
Do anything else necessary to enable them to carry out their duties.	Like what?
Issue an improvement notice.	What would it be for?
Issue a prohibition notice.	What would it be for?
Initiate prosecutions.	Against the company or a director?
Seize, destroy or render harmless any article or substance which is a source of imminent danger.	What would I seize, destroy or render harmless?

EXAMPLE 3: GLOSSARY OF FIRE

Writing your own glossaries can be a useful technique, especially when having to learn technical terms.

The following is a list of keywords that are used in fire safety. Note: this list is not exhaustive.

Fire term	*Definition*
Flashpoint	The lowest temperature at which the heat from the combustion of the burning vapour is capable of producing more vapour that, once ignited, is able to sustain the combustion cycle.
Fire point	The lowest temperature at which a substance will produce sufficient vapour to flash across its surface momentarily when a flame is applied.
Auto-ignition temperature	The lowest temperature at which a substance will ignite spontaneously without any other ignition source.
Vapour density	The density of a gas, expressed as the mass of a given volume of the gas divided by the mass of an equal volume of a reference gas (such as hydrogen or air) at the same temperature and pressure.
Vapour pressure	The pressure of a vapour given off by (evaporated from) a liquid or solid, caused by atoms or molecules continuously escaping from its surface. Vapour pressure is measured in units of pressure such as Pascal's.
Flammable	Liquids which have a flashpoint below 55°C but which are not highly flammable.
Highly flammable	Liquids which have a flashpoint below 21°C but which are not extremely flammable.
Extremely flammable	Liquids that have a flashpoint lower than 0°C.
Upper flammable limit	The maximum concentration of vapour in the air above which the propagation of flame will not occur in the presence of an ignition source.
Lower flammable limit	The minimum concentration of vapour in the air below which the propagation of flame will not occur in the presence of an ignition source. Also referred to as the lower explosive limit.
Class A	Fires that involve solid materials, usually of an organic nature, such as wood, cardboard, paper, hardboard, and soft furnishings such as carpets and curtains, in which combustion normally takes place with the formation of glowing embers.

Class B	Fires that involve liquids such as petrol, paraffin, white spirit, thinners, varnish and paints, or liquefiable solids such as candles (wax) and fats.
Class C	Fires that involve gases such as LPG (e.g. butane, propane) or those involving natural gas.
Class D	Fires that involve metals such as sodium, lithium, manganese and aluminium when in the form of swarf or powder.
Class F	Fires that involve cooking mediums such as vegetable or animal oil and fats in cooking appliances.
Flashover	In relation to the behaviour of fire in buildings and other enclosed spaces, flashover is the point at which the whole room or enclosure where the fire started becomes totally engulfed in fire.
Backdraft	An explosive reaction that occurs when a fire occurs in an enclosed space, which has died down due to insufficient oxygen, and then is suddenly provided with large quantities of oxygen.
Deflagration	A combustion wave propagating from an explosion at subsonic velocity relative to the unburnt gas immediately ahead of the flame (flame front).
Detonation	A combustion wave propagating from an explosion at supersonic velocity relative to the unburnt gas immediately ahead of the flame (flame front).
BLEVE	Boiling Liquid Expanding Vapour Explosion – an explosion due to the flashing of liquids when a vessel with a high vapour pressure substance fails.
CGE	Confined Gas Explosion – explosions within tanks, process equipment, sewage systems, underground installations, closed rooms, etc.
UVCE	Unconfined Vapour Cloud Explosion – a vapour/gas explosion (deflagration or detonation) in an unconfined, unobstructed cloud.
Conduction	The transfer of heat through solid materials involving the molecule-to-molecule transfer of heat through conducting solids such as metal beams or pipes to other parts of the building and igniting combustible or flammable materials.
Radiation	Involves the emission of heat in the form of infrared radiation, which can raise temperatures of adjacent materials; for example, electric fire elements.
Convection	The transfer of heat from a liquid or gas (i.e. air, flames or fire products) to a solid or liquid surface. Heat can be carried by rising air currents (convection) to cause a buildup of hot gases.
Direct burning	When the flames and/or heat reach the combustible materials and ignite them.

ACTIVITY

What other areas of health and safety could you develop a glossary for?

___

___

___

Write a glossary during your health and safety course. If possible, group together related items rather than alphabetically.

EXAMPLE 4: MATERIAL SAFETY DATA SHEETS

The following 16 headings are found on all material safety data sheets:

1. Identification of the substance/mixture and of the company.
2. Hazards identification.
3. Composition/information on ingredients.
4. First aid measures.
5. Fire-fighting measures.
6. Accidental release measures.
7. Handling and storage.
8. Exposure controls/personal protection.
9. Physical and chemical properties.
10. Stability and reactivity.
11. Toxicological information.
12. Ecological information.
13. Disposal considerations.
14. Transport information.
15. Regulatory information.
16. Other information.

ACTIVITY

Download a safety data sheet of your choice and review the headings, looking at what is included under each heading.

EXAMPLE 5: LEARNING THE PRINCIPLES OF RISK ASSESSMENT 1

Many health and safety courses use the following five steps to risk assessment:

Step 1	Identify the significant hazards.
Step 2	Decide who may be harmed and how.
Step 3	Evaluate risks and decide on precautions.
Step 4	Record your findings and implement them.
Step 5	Review the assessment and update if necessary.

KEY NOTE

By working through each of the steps and thinking about what each would actually mean for your workplace, you will increase your understanding and therefore be more likely to remember the steps.

The following table guides you through this process.

Step 1	Consider activities that are being undertaken in your workplace and identify the significant hazards involved. This may be done by: ■ Simply observing what may be happening. ■ Reviewing accident data. ■ Looking at manufacturers' instructions. ■ Talking to experienced personnel.
Step 2	Identify those exposed to hazards, including both employees and non-employees: in particular, groups who may be especially at risk; for example, young persons, pregnant/nursing mothers, and those whose first language is not that used at the place of work, etc.
Step 3	Evaluation of risks arising from identified hazards. By calculating the 'likelihood' of an action occurring and its 'severity', this result gives the risk rating showing a priority for action and may indicate the type of control measures needed and their priority.
Step 4	Recording 'significant findings' (i.e. ignoring the trivial ones). The information must be made available to all those involved in the task or activity.

Step 5	Review and revise the assessment at regular intervals or more frequently if there are: ■ Changes in the processes, work methods or materials used. ■ Introduction of new, or the modification of, existing plant. ■ Availability of new information on hazards and risks. ■ Availability of new or improved control measures or techniques. ■ A change in legislation. ■ Changes in personnel; for example, the employment of young or disabled persons. ■ After the passage of time, for example, one year.

The above may indicate that the original assessment is no longer valid.

EXAMPLE 6: LEARNING THE PRINCIPLES OF A FIRE RISK ASSESSMENT

Many health and safety courses use the five steps to fire risk assessment:

Step 1	Identify fire hazards.
Step 2	Identify people at risk.
Step 3	Evaluate, remove, reduce and protect from risk.
Step 4	Record, plan, inform, instruct and train.
Step 5	Review.

KEY NOTE

By working through each of the steps and thinking about what each step would actually mean for your workplace, you will increase your understanding and therefore you will be more likely to remember the steps.

The following table guides you through this process.

Step 1	Identify fire hazards: involves looking at the workplace, having an understanding of work carried out and identifying sources of: ■ Oxygen. ■ Ignition. ■ Fuel. Special note should be made of: ■ Electrical sources of ignition. ■ Smoking arrangements. ■ Arson risk. ■ Portable heating appliances. ■ Cooking facilities. ■ Lighting (risk of ignition). ■ Housekeeping. ■ Contractors/others in the workplace. ■ Dangerous substances – having an explosive/flammable risk.
Step 2	Identify people at risk and review fire protection measures to ensure safety of: ■ Employees. ■ Non-employees. ■ Disabled/young persons, etc. Consideration/evaluation of the following needs to be addressed: ■ Means of escape – number of exits/routes blocked? ■ How fire would spread/what will limit it spreading? ■ Lighting – is it suitable? ■ Fire signage – is it suitable? ■ Warning/alarms – are they suitable? ■ Number and types of fire extinguishing appliances used. ■ Automatic sprinkler system – is one needed?
Step 3	Evaluate, remove or reduce and protect from risk: this step involves addressing issues raised in Step 2 and ensuring delegation of responsibility for fire safety. The following procedure is also required: ■ Appointment of fire marshals/wardens. ■ Testing of alarm system (weekly). ■ Development of an action plan with outstanding issues that need to be dealt with.
Step 4	Record, plan, inform, instruct and train. Details of risk assessment should be available for all employees/ others in the workplace. This information needs to be supported by carrying out drills and fire training suitable to the workplace and its activities. Records need to be kept of: ■ Fire drills. ■ Fire training. ■ Alarm testing. ■ Emergency lighting.

Step 5	Review risk assessment. A review should be carried out annually, or if there is a change in activity or premises that will have rendered the assessment invalid, or in the event of a fire!

KEY NOTE

By looking at both the general information and the five steps to fire risk assessment, you will be able to compare and contrast the similarities and differences.

EXAMPLE 7: SIGNS IN HEALTH AND SAFETY

By not participating in a training session (i.e. just listening to the trainer, reading the book and watching PowerPoint slides), you will most likely only remember half of what has been covered. If you make notes this will increase your recall by one-fifth! To remember even more you will need to engage in some form of task. This may be as simple as writing out an answer to a question and then checking the answer.

You can apply this principle to almost any area of health and safety; for example, if you are trying to remember the different types of safety sign you will need first to be able to know the colour and shape.

Sign	*Colour*
Safe condition	Green and white
Mandatory	Blue and white
Prohibition	Red and white
Hazard	Yellow and black

Sign	*Shape*
Safe condition	Rectangle
Mandatory	Circle
Prohibition	Circle
Hazard	Triangle

Now that you have read the information you will need to see an example for each sign.

Sign	*Example*
Safe condition	First aid
Mandatory	Eye protection
Prohibition	No smoking
Hazard	Live cables

The final stage in this example is to write out the colour and shape plus an alternative example (can you think of one from your place of work?)

ACTIVITY

Sign	Colour and shape	Alternative example
Safe condition	________________	________
Mandatory	________________	________
Prohibition	________________	________
Hazard	________________	________

KEY NOTE

If you have looked up an example (i.e. you have performed a task), you are more likely to remember the information.

If you need to know the content of a safety policy (i.e. the three sections: Statement of Intent, organisation, and arrangements, and the possible headings of the arrangements), you could:

- Read information about the content in a health and safety course book.
- Listen to someone describe what should be included.
- See a good example of a health and safety policy with its arrangements.
- Write your own version of a health and safety policy after seeing a couple of examples.

KEY NOTE

As you will appreciate, the more work you do the more you will remember. This also applies to sitting past exam questions based on the learning outcomes you have just undertaken.

During a training session your trainer is likely to use PowerPoint to reinforce what he or she is saying. As the student you will then hear and see the information.

After the training session students could simply read their notes, but if they rewrite their notes and then tell someone else about what they have learned they will probably retain more information. By sitting a mock exam they will be increasing retention of the information. Each stage reinforces the learning that has already taken place.

By using as many of these stages as possible you are more likely to remember the information needed for your exam.

EXAMPLE 8: EXAMPLE OF NOTICES (HSE)

Health and safety inspectors under the Health and Safety at Work etc. Act 1974, either for HSE or for a local authority, have the authorization to:

- Issue an improvement notice.
- Issue a prohibition notice.

Examples may be found at http://www.shponline.co.uk/legislation-and-enforcement:

IMPROVEMENT NOTICE

Judge slams chemical firm's 'abysmal incompetence'
31 July 2012

A Welsh chemical company has been fined £100,000 for failing to comply with three Improvement Notices. Mold Crown Court heard Euticals Ltd (formerly Archimica Ltd) manufactures and distributes specialty chemical products at its facility in Sandycroft, Deeside. During a series of HSE visits to the site between March 2010 and

October 2010, the company failed to demonstrate that it had suitable measures in place to prevent a major incident. On 29 March 2010, HSE issued an Improvement Notice requesting the company supply missing information from the safety report of its premises. During subsequent visits HSE Inspector Mark Burton identified that the firm hadn't assessed the roles and responsibilities of staff for the management of major hazards. On 1 November 2010 he issued a second Improvement Notice, which required the firm to put arrangements in place to manage major accident hazards. Inspector Burton issued another Improvement Notice on the same day, instructing Euticals to put control measures in place to ensure that intrinsic safety equipment was inspected every three years. The company had failed to carry out inspections on a large amount of this equipment for more than six years and 177 items had failed inspections without any remedial action being taken. HSE extended the period of compliance for each notice multiple times, but the company failed to take the necessary actions. Inspector Burton said: 'The company had plenty of opportunity to comply with the Improvement Notices after repeated visits from HSE and they still chose not to. They deal with dangerous chemicals every day and have a legal responsibility to make sure that how they do that is safe. That responsibility extends not only to their employees but to the site's neighbours and any visitors, as well as the environment. There's the possibility for a major incident when manufacturing and distributing this kind of product and failing to plan for it could have devastating consequences.' On 27 July, Euticals pleaded guilty to breaching s33(1)(g) of the HSWA 1974, for failing to comply with the notices. In addition to the fine, it was ordered to pay £8344 in costs. In mitigation, the company said it has now spent more than £1million to comply with the notices. Delivering his sentence, Judge Philip Hughes said: 'An aggravating feature is the defendant company's reckless regard for adhering to the law and somewhat dismissive attitude to those in HSE trying to guide them and neglecting to take preventative measures to reduce the risks. This is a case, which has demonstrated in the defendant company, a persistent lack of management control and abysmal level of disorganization and incompetence.'

http://www.shponline.co.uk/incourt-content/full/judge-slams-chemical-firms-abysmal-incompetence

PROHIBITION NOTICE

Prohibition Notice breached during follow-up inspection
6 September 2011

An HSE inspector has slammed two construction companies for 'appalling' safety standards at a building site in London. The safety watchdog prosecuted principal contractor, Kubik Homes Ltd, and sub-contractor, Bellway Developments Ltd, after inspectors made several visits to the site in Wimbledon. Inspectors found that there was no safe access to the first floor of the building and a makeshift 'staircase' had been formed using a bag of sand and piles of blocks. In order to access the roof of a hut, which was positioned on the first floor, workers walked up wooden planks that had no edge protection to prevent falls. Inspectors discovered a 2.5 metre-deep excavation, which had no barriers in place to stop workers from falling into the pit, and no supports in place to prevent the excavation from collapsing. The site was littered with multiple slip and trip hazards, uneven land, and building materials that had been stacked excessively high and stored haphazardly. The toilet at the site was also found to be in a filthy condition and had a leaking cold-water supply. On 7 October 2009, Kubik Homes was issued with several Prohibition Notices, one of which was breached while inspectors were on site during a subsequent visit. The notices required: work to stop at the site until slip and trip hazards had been removed; suitable edge protection to be put in place; a safe method of access to the first floor to be created, supports to be implemented to secure the sides of the excavation; and a competent person to be put in place to manage health and safety at the site. When inspectors returned to the site they saw two men walking on the first floor, which accounted to a breach of one of the Prohibition Notices. It became clear that work was still continuing in an unsafe manner, so a further Prohibition Notice was issued, which required all work to stop until competent site management was put in place. HSE Inspector Loraine Charles explained that the whole site was badly managed and Bellway Developments Ltd had failed to properly plan work at the site. The company was issued an Improvement Notice on 28 October 2009 for failure to provide sufficient information, instruction, training and supervision for the director and site supervisor, and for failing to protect its workers.

Inspector Charles also revealed that none of the workers from either company had a recognized qualification in site management. She said: 'Although there was no incident, the potential danger to the workers was very high. Conditions on this site were simply appalling. This is a shocking example of bad management of a construction site and it is a miracle that no one was injured. Both these companies failed to understand the nature of their duties under health and safety law and failed to sufficiently improve conditions on the site despite repeated interventions by the HSE.' Kubik Homes Ltd appeared at City of London Magistrates' Court on 30 August and pleaded guilty to breaching s3(1) of the HSWA 1974. It was fined £8000 and ordered to pay £2426 in costs. Bellway Developments Ltd also pleaded guilty to breaching s3(1) of the HSWA at the same hearing and was fined £8000, plus costs of £2384.

http://www.shponline.co.uk/incourt-content/full/prohibition-notice-breached-during-follow-up-inspection

ACTIVELY READING

Trying to read a course book, e-book, webpage or pdf document without understanding what information you need to receive, remember and then recall can be difficult. Many hours are lost by health and safety students passively reading or listening to/watching podcasts and videos without focusing on what they need to receive and remember.

The most important step is to ensure that you have a copy of the syllabus with you.

CASE STUDY

Sheila (a distance learning student for the NEBOSH General Certificate) printed out the pdfs sent by her course provider. While keeping a copy of the syllabus by her side, Sheila quickly read through all of the material first without making notes, only highlighting some of the text. Then she went back to the beginning and reread each section, making notes in her notebook.

If you are reading something that is not within the scope of one of the learning outcomes, you are simply wasting your study time!

CASE STUDY

Steve copied out each of the learning outcomes from the syllabus on to the top of new A4 pages in his spiral notebook and then made notes from the e-learning programme supplied by his course provider.

There are many sources of information available for you to read, listen to and watch.

Some students find it useful to read around the subject they are studying; for example, when studying a particular topic try to find the relevant HSE documents from their website: http://www.hse.gov.uk/pubns/hsebooks-catalogue.pdf.

CASE STUDY

Steve was given a copy of a NEBOSH course book when he was studying for the CIEH Level 4 course. The course trainer used the CIEH book in class, but Steve was able to use the NEBOSH book to look up specific areas that he wanted to read about in more detail.

KEY NOTE

Remember to read the most up-to-date texts – always check you have the latest edition of a book!

ASSESSING YOUR LEARNING

Your trainer will use a mixture of questions to ensure that the information has been assimilated:

- A closed question requires a short, normally yes-or-no answer – this gives you the opportunity to simply say yes or no.
- An open question gives you the opportunity to expand your answer.

There are ranges of different assessment methods available. The following looks at initial, formative and summative assessments:

- **Initial assessment:** before you start the course see how much you already know. You may be able to read the course book and realize you know some or very little of the material. If you attempt some example exam questions this will help you guage how much you need to learn.
- **Formative assessment:** assess yourself throughout the learning process; for example, at the end of a section attempt to answer some example exam questions from that section and ask your tutor to mark them for you. *Introduction to Health and Safety at Work*, the handbook for the NEBOSH National General Certificate, includes questions that may be used to check your understanding as you proceed through the course.
- **Summative assessments:** assessments that are carried out at the end of a course. You can test your overall understanding of the material taught by sitting a full mock exam and asking your tutor to mark it for you.

CASE STUDY

Richard asked his tutors on the NEBOSH Certificate course to mark several mock exams. This was done for each of the two units. His tutors complied with his request and he achieved a Credit as a result.

Feedback is a valuable tool; your trainer should be able to provide constructive feedback. This may be for activities undertaken during the course or independent study. If you sit a mock/past question exam and your trainer says it was only worth four marks out of eight without making any comments this only provides you with the information that you achieved 50 per cent. You will also need to know why you lost marks and how you could improve your answer.

For example:

- Did you write down a list rather than an outline?
- Did you misread part of the question; for example, did you provide the hazards when the question asked you for the controls?
- Was some of your handwriting illegible?
- Did you repeat your answers, using different words?
- Did you spend too much time writing about the same point?

ACTIVELY REVISING WITH LEARNING OUTCOMES

The learning outcomes may be used as a great revision tool. By making notes as soon as possible after class, you will be able to remember more about each learning outcome.

Attempt the following, using the notes made after class:

ACTIVITY

- Cover your notes and read the learning outcome out loud.
- Try to remember as much as you can (just think about it).
- Now uncover your notes and read through them – what did you remember, what did you miss?

Keep repeating the above for all the learning outcomes.

Since exam questions are based on learning outcomes, by repeating this activity you are not only covering the content of the course in full, but, in essence, you are also testing yourself on the types of questions that may come up.

REVISION TECHNIQUES

There are many revision techniques that may be used. It is not about memorizing lists, even though this is sometimes useful, but it is more about understanding the material. Some students find it difficult to memorize information, but when they understand how it links together it becomes easier.

Revision should be ongoing, starting after day one of the course, reviewing your notes made in class and those given on the course, which is likely to be the course book and PowerPoint slides.

Depending upon how much time you have before sitting the exam it is well worth taking a short break from study and revision and doing something completely different and relaxing. Then, once refreshed, review *all* your notes and course book, going through the material as quickly as possible to give yourself a full overview of the course. You will see how different parts/elements are linked. Once you have perused all the material it is now time to focus on the areas that you need to spend more time revising.

CASE STUDY

John: 'When I am about to start the revision for any course I read over all the notes in the course book – as I have done this, memories come flooding back of things said in class, and examples given by the trainer and other students.'

KEY NOTE

By reading the same information presented in a different way/format/style you will begin to remember; for example, your thoughts may be 'In the course book it said x y z – but in my workplace policy it expands these further!'

KEY NOTE

Throughout this book I have repeated key health and safety information; for example, five steps to risk assessment. As you reread these steps in this book your brain will be making connections that will help you remember.

KEY HEALTH AND SAFETY TOPICS

This section provides students with both:

- Key information for their studies.
- Examples of activities that could help them remember specific topics within a health and safety course.

> **KEY NOTE**
>
> Health and safety topics covered in brief are listed below. These topics have been selected as they are common on most health and safety courses. By reading them here it will either give you the advantage of knowing many of the topics that will come up on your course; or if you have already sat the course/are sitting it, then this section is an excellent opportunity for revision.

- Accidents/incidents.
- Health hazards.
- Economic benefits.
- Sources of information.
- Roles and responsibilities in health and safety – UK.
- Roles and responsibilities in health and safety – International.
- Consulting and informing.
- Health and safety management system.
- Health and safety policy.
- Monitoring.
- Accidents and ill-health.
- Immediate and underlying causes of an accident.
- Recording and reporting.
- Methods of improving health and safety performance.
- Risk assessment.
- Health surveillance.
- Work equipment.
- Vehicles in the workplace – transport and pedestrians.
- Electricity.
- Fire.
- Ergonomic factors – manual handling.
- Ergonomic factors – repetitive physical activities/rapid tasks.

- Ergonomic factors – display screen equipment.
- Hazardous substances.
- Routes of entry of hazardous substances into the body.
- Welfare and the work environment.
- Stress.
- Violence.
- Noise.
- Vibration.
- First aid.

The activities below represent only a small example of possible activities. Once you have reviewed the following it will be possible to develop your own activities. The following table does not aim to cover all aspects of a health and safety course, but to summarize the main areas that may be covered.

The information used to develop the following has been taken from various HSE industrial guidance (INDG) documents, including:

Management and policy	INDG355, INDG275, INDG259, INDG132, INDG 232, INDG 213
Consulting and informing	INDG232
Health and safety management system	INDG275
Risk assessment	INDG163
Health surveillance	INDG304
Work equipment	INDG229, INDG270
Vehicles in the workplace – transport and pedestrians	INDG199, INDG255, INDG413
Electricity	INDG231, INDG236,
Fire	INDG227
Ergonomic factors – manual handling	INDG145, INDG242
Ergonomic factors – repetitive physical activities/rapid tasks	INDG171
Ergonomic factors – display screen equipment	INDG36
Hazardous substances	INDG136, INDG350, INDG352
Welfare and the work environment	INDG244
Stress	INDG424
Violence	INDG69
Noise	INDG362, INDG363
Vibration	INDG175, INDG296
First aid	INDG214

ACTIVITY

Once you have read the subject matter below look up the relevant HSE INDG documents. These may be found at http://www.hse.gov.uk/pubns/ and follow the links.

KEY NOTE

The following activities may also be used in class by trainers when delivering health and safety qualifications.

ACCIDENTS/INCIDENTS

The top three causes of fatalities, major injuries and absences over three days are:

Top three causes of fatalities:	– Fall from a height. – Struck by moving vehicle. – Struck by moving – including flying/falling object.
Top three causes of major injuries:	– Slips, trips and falls on the same level. – Fall from a height. – Injured while handling, lifting or carrying.
Top three causes of absences over three days:	– Injured while handling, lifting or carrying. – Slips, trips and falls on the same level. – Struck by moving – including flying/falling object.

ACTIVITY

Consider which of the above incidents are most likely to occur in your workplace.

HEALTH HAZARDS

There are five categories of health-related hazards:

Category	*Impact*
1. Physical	Hearing loss. Hand/arm vibration.
2. Chemical	Chemical agents that affect the body; for example, skin – dermatitis/lungs – asthma.
3. Biological	Biological agents that affect the body; for example, blood-borne viruses (e.g. HIV).
4. Physiological/psychological	Stress and mental health.
5. Ergonomic	Muscular skeletal disorders.

ACTIVITY

Consider which of the above hazards would be likely to occur and to have the greatest consequence within your workplace.

ECONOMIC BENEFITS

The economic benefits for an organization having good health and safety standards are:

- Improved company reputation.
- Improved production.
- Improved staff morale.
- Increased levels of compliance with rules and procedures.
- Reduced absenteeism.
- Fewer accidents.
- Reduced damage to equipment.
- Fewer fines and compensation claims.
- Reduced ill-health.
- Reduced insurance premiums.
- Fewer staff complaints.
- Reduced staff turnover.

ACTIVITY

Which are the most significant economic benefits for your workplace and why?

Good health and safety makes financial sense! In the event of an accident there are various costs that can result – direct and indirect – and which may be insured or uninsured.

Direct costs

Costs incurred by an organization or individual as a result of an accident or incident that can be readily accounted for; for example, fines, wage compensation, repair to plant or equipment, etc.

Indirect costs

Costs incurred by an organization or individual as a result of an accident or incident that can be inferred but which do not take the form of direct monetary outlays; for example, disruption to the work of other employees.

Insured costs

Financial losses resulting from a workplace accident that are covered by an insurance policy; for example, ill-health, injury and damage.

Uninsured costs

Financial losses resulting from a workplace accident that are not covered by an insurance policy; for example, lost time, extra wages, fines, legal costs, etc.

ACTIVITY

List the costs incurred if a serious accident happened in your workplace.

Direct costs

Indirect costs

Insured costs

Uninsured costs

SOURCES OF INFORMATION

Sources of information used by an organization for health and safety purposes may be internal or external in nature. An organization should review all aspects of health and safety, and look to ways of improving its health and safety performance regularly.

Internal information:

- Accident and incident data.
- Accident investigation reports.
- Risk assessments.
- Safe systems of work.
- Emergency procedures.
- Health and safety monitoring.
- Health surveillance.
- Insurance company reports.
- Internal health and safety policy.
- Near-miss data.
- Maintenance records.
- Health and safety surveys.
- Training records.
- Safety committee minutes.

External information:

- Material Safety Data sheets.

- Manufacturers' information.
- Trade Associations.
- Awarding Bodies.
- National/International standards.

ACTIVITY

Review the internal information within your company and identify useful sources of external information (make a list of relevant websites).

ROLES AND RESPONSIBILITIES IN HEALTH AND SAFETY – UK

Under health and safety legislation in UK the main employer's duties are as follows:

Health and Safety at Work etc. Act 1974

- To ensure, so far as is reasonably practicable, the health, safety and welfare at work of all employees.
- To prepare a written health and safety policy.

Management of Health and Safety at Work Regulations 1999

- To carry out a risk assessment, covering both workers and others who may be affected by their work or business.
- To make appropriate arrangements for the effective planning, organization, control, monitoring and review of protective and preventative measures.
- To appoint one or more competent persons to assist in managing risk.

KEY NOTE

Both the Act and Regulation have more details that you would be expected to know if you are taking a National Qualification. By looking these up; writing them out you are more likely to remember the information than just reading it once.

According to health and safety legislation in the UK the employees' main duties are as follows:

Health and Safety at Work etc. Act 1974

- To take reasonable care of their own health and safety and that of others who may be affected by what they do or do not do.
- To cooperate with their employer on health and safety issues.
- Not interfere with, or misuse, anything provided for their health, safety or welfare.

Management of Health and Safety at Work Regulations 1999

- Correctly using work items provided, including personal protective equipment, in accordance with training or instructions.
- To inform their employer of any situation that represents a serious and immediate danger or shortcomings in arrangements.

ACTIVITY

Consider how this legislation is applied in your own workplace – look up both pieces of legislation and review the other duties within them.

ROLES AND RESPONSIBILITIES IN HEALTH AND SAFETY – INTERNATIONAL

International Labour Organization (ILO)

Member states of the United Nations comply with the various conventions and recommendations set by the International Labour Organization (ILO). The main convention and recommendation for occupational health and safety are:

- Occupational Safety and Health Convention (C155) – a goal-setting policy for companies and nations.

- Occupational Safety and Health Recommendation 1981 (R164) – supplements C155 and gives more guidance on how to comply with its policies.

Employers' duties

Article 16 of C155 identifies duties placed upon employers:

- To provide and maintain workplaces, machinery, equipment and work processes.
- To ensure that chemical, physical and biological substances and agents are without risk to health when protective measures have been taken.
- To provide adequate protective clothing and equipment to prevent risks of accidents or adverse health effects.

Article 19 of C155 states that every worker must be:

- Given adequate information on actions the employer has taken to ensure safety and health.
- Given the right to the necessary training in safety and health.
- Consulted by the employer on all matters of safety and health relating to their work.
- Given the right to leave a workplace a worker has reason to believe presents an imminent and serious danger to life or health, and not be compelled to return until it is safe.

Article 10 of R164 identifies additional obligations placed on employers to:

- Provide and maintain workplaces, machinery and equipment, and to use working methods that are safe.
- Give necessary instruction, training and supervision in the application and use of health and safety measures.
- Introduce organizational arrangements relevant to activities and size of undertaking.
- Provide PPE and clothing without charge to workers.
- Ensure that work organization, particularly working hours and rest breaks, does not adversely affect occupational safety and health.

- Take reasonably practical measures with a view to eliminating excessive physical and mental fatigue.
- Keep abreast of scientific and technical knowledge to comply with the above.

Workers' duties

Article 19 of C155 also places duties on workers, expanded in R164 as follows:

- Take reasonable care of their own safety and that of other people.
- Comply with safety instructions.
- Use all safety equipment properly.
- Report any situation which they believe could be a hazard and which they cannot themselves correct.
- Report any work-related accident/ill-health.

> **ACTIVITY**
>
> Download both documents from the International Labour Organization (ILO) website http://www.ilo.org.
>
> Read both documents and compare and contrast the similarities and differences to UK health and safety legislation.

CONSULTING AND INFORMING

The difference between consulting and informing is as follows:

Consulting

A two-way process in which the employer listens to, and takes account of, the views of workers before a decision is made.

Informing

A one-way process; for example, providing information to workers about hazards, risks and control measures.

There are many benefits of consulting with employees/workers. It provides:

- Access to the experience of a range of people in the organization (including employees) who have extensive knowledge of their own job and the organization.
- Better decisions about health and safety are made – because they are based on the input of those who create and understand the nature and extent of the risk(s).
- Greater cooperation and trust – because employers and employees talk to each other, listen to each other and gain a better understanding of each other's perspectives and hence views.
- Healthier and safer workplaces – because employee input is valuable to identify hazards, assess risks and develop ways to control or remove risks.
- Stronger commitment to the implementation of decisions or actions – because employees have been actively involved in reaching these decisions.

> **ACTIVITY**
>
> Identify the ways in which consulting and informing is undertaken in your workplace – is it effective?

HEALTH AND SAFETY MANAGEMENT SYSTEM

Requirements for planning and organizing the management of health and safety in the workplace within an organization can be undertaken by using a formal management system; for example, 'Successful Health and Safety Management' (HSG65).

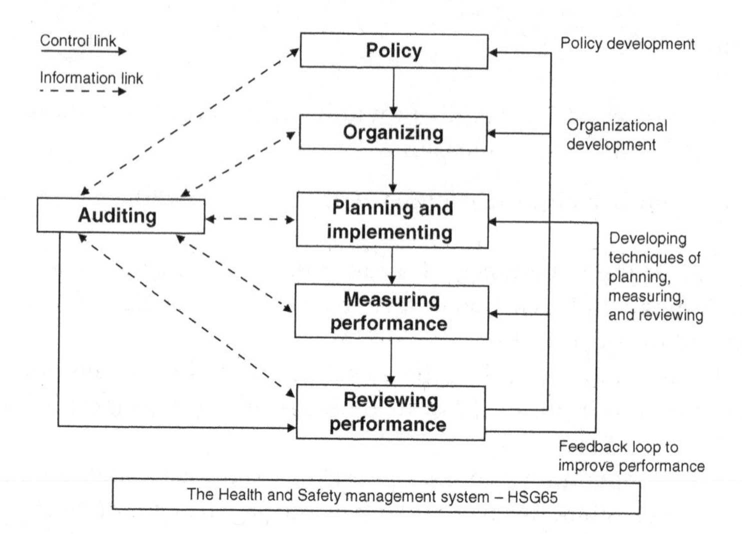

The Health and Safety management system – HSG65

Developing a safety management system starts with establishing:

1. **Policy** which states management's intentions, and sets clear aims, objectives and targets.
2. **Organizing** for health and safety by allocating responsibilities, and establishing effective communication and commitment at all levels.
3. **Planning and implementing** is where practical plans are developed to meet the objectives and effective control measures introduced based on risk assessment, which will include addressing the health and well-being of the workforce.
4. **Measuring performance** will involve proactive and reactive monitoring systems. These should be introduced to evaluate the performance against the objectives and targets, and to consider options for improvement and to reset targets.
5. **Reviewing performance** would involve addressing the results from proactive and reactive monitoring.
6. **Auditing** of all parts of the safety management system, in a systematic way, would assess compliance with health and safety management procedures and identify where existing standards are inadequate or deficient.

ACTIVITY

Review your workplace management system – does it follow this model?

HEALTH AND SAFETY POLICY

The three parts of a health and safety policy are:

- Statement of Intent.
- Organization.
- Arrangements.

Statement of Intent

The statement of intent is a one-page document that sets out the aims and objectives of the organization. This demonstrates management's commitment to health and safety, and sets the goals, targets and objectives for the organization. It commits the employer to a standard at least as high as that required by health and safety legislation. The statement of intent may include targets set to improve the health and safety of the organization.

Companies should set health and safety targets for a number of reasons, which include:

- To give evidence of management commitment.
- To prioritize and focus on important issues.
- To motivate staff by providing them with something tangible to aim for and to encourage their ownership of health and safety issues.
- To enable performance with standards to be measured and to identify improvements that have been made.
- To enable trends to be identified.
- To meet standards of safety management systems.
- To highlight the important role targets can play in facilitating the measurements and review of performance.

ACTIVITY

Write a revised statement of intent for your workplace.

OR

Review several statements of intent using Google search on the Internet.

Organization

Organization outlines the roles and responsibilities within a company. Often an organogram is used to show the lines of authority:

- To allocate health and safety responsibilities within the company.
- To ensure effective delegation and to set up lines of communication.

> **ACTIVITY**
>
> Draw an organogram for your workplace.
>
> OR
>
> Review your workplace's organogram – is it correct?

Arrangements

Arrangements set out in detail the specific systems and procedures that aim to assist in the implementation of the statement of intent. The following list is an example of some of the specific arrangements that might be implemented:

Premises	■ Signage. ■ Smoking. ■ Workplace environmental factors: lighting, humidity, ventilation, thermal, space, welfare and hygiene. ■ Obstructions/slips, trips and falls. ■ Services (gas, water and electricity).
Plant and equipment	■ Work equipment. ■ Maintenance. ■ Risk assessments. ■ Noise and vibration. ■ Radiation. ■ Substances hazardous to health (no biological agents present). ■ Pressure systems. ■ Mechanical handling.

People	■ Information, instruction, training and supervision. ■ Information/notices/insurance/law. ■ Consultation with employees. ■ Stress. ■ Violence. ■ Alcohol and drugs. ■ PPE. ■ Lone working. ■ Health surveillance. ■ Air monitoring. ■ Contractors and consultants.
Procedures	■ Fire, first aid and other emergencies. ■ Electrical. ■ Manual handling. ■ Monitoring.

ACTIVITY

Can you identify any further arrangement headings?

Reviewing health and safety policies

Reasons for reviewing a health and safety policy include:

- Changes in the processes, work methods or materials used.
- Introduction of new, or the modification of existing, plant.
- Availability of new information on hazards and risks.
- Availability of new, or improved, control measures or techniques.
- A change in legislation.
- Changes in personnel; for example, the employment of young or disabled persons.
- After the passage of time; for example, one year.

ACTIVITY

Find out when, who by and why your workplace health and safety policy was updated.

MONITORING

Types of monitoring are:

- Active monitoring.
- Reactive monitoring.

Active monitoring

Active monitoring is proactive. By examining different aspects of the organization's health and safety performance it is possible to identify areas for improvement.

Common methods of active monitoring include:

Benchmarking	A planned process by which an organization compares its health and safety processes and performance with others to learn how to: reduce accidents and ill-health; improve compliance with health and safety law; and/or cut compliance costs.
Health surveillance	The specific health examination at predetermined frequencies of those at risk of developing further ill-health or disability and those actually or potentially at risk by virtue of the type of work they undertake during their employment; for example, involving asbestos, lead, etc.
Inspections	Regular/scheduled activities identifying existing conditions and comparing them with agreed performance objectives. May be observations of people, procedures, premises, plant and equipment.
Safety surveys	A detailed examination of a number of critical areas of operation or an in-depth study of the entire health and safety operation of premises; for example, focuses on specific aspects of work – noise, manual handling, etc.
Sampling	An organized system of regular random sampling, the purpose of which is to obtain a measure of safety attitudes and possible sources of accidents through the systematic recording of hazard situations observed during inspections made along a predetermined route in the workplace – for example, the examination of the guards on one or two machines – will give an indication as to the state of all guards on all machines.
Tours	Carried out by management to address the effectiveness of risk control measures. In most cases these will be unscheduled.

Audits

A safety audit is a systemic critical examination of an organization's health and safety management system, involving a structured process including the use of a series of questions and the examination of documentation. The purpose is to collect independent information with the aim of assessing the effectiveness and reliability of the system and suggesting corrective action when this is thought to be necessary. It is carried out by trained auditors, who may be internal or external to the organization.

The findings of a safety audit may be used in a number of ways to improve health and safety performance, such as:

- Assisting with the allocation and prioritization of resources.
- By means of subsequent audits at regular intervals, assisting in the continual improvement of the management system.
- Communicating its findings to management and staff.
- Enabling comparison and benchmarking with other similar organizations.
- Enabling remedial action.
- Giving an indication of the organization's commitment to health and safety.
- Identifying compliance and non-compliance and the reasons for the latter, thus informing management of actions to be taken.
- Identifying strengths and weaknesses in the management system.

Reactive monitoring

Reactive monitoring is concerned with reviewing events that have occurred in order to learn from any mistakes made. It establishes what systems and procedures can and/or should be put in place to prevent a reocurrence. These include:

- Analysis of absences/ill-health data.
- Investigation of the actions taken by the enforcement authorities.
- Investigation of complaints by the workforce.
- Investigation into property damage, dangerous occurrences, near misses and accidents.

For example, accident data may be used to improve health and safety performance:

- Identify specific problem areas by recording instances where control measures have failed.
- Accident data can also help to raise awareness in the minds of both managers and employees of health and safety in general, and of specific problems in particular.

ACTIVITY

Discuss with a work colleague the types of monitoring that are carried out in your workplace. Can you think of any additional ones that are not used at present in your workplace?

ACCIDENTS AND ILL-HEALTH

Common workplace accidents include:

- Contact with moving machinery or materials being machined.
- Struck by a moving, flying or falling object.
- Hit by a moving vehicle.
- Struck against something fixed or stationary.
- Injured while handling, lifting or carrying an object.
- Slips, trips and falls on the same level.
- Falls from a height.
- Trapped by something collapsing.
- Drowned or asphyxiated.
- Exposed to, or in contact with, a harmful substance.
- Exposed to fire or explosion.
- Contact with electricity or an electrical discharge.
- Injured by an animal.

Common causes of work-related ill-health include:

- Ergonomic: muscular skeletal disorders (back and upper limbs or neck).
- Physiological: stress, depression and/or anxiety.
- Respiratory: breathing or lung problems.
- Diseases: infectious and skin.
- Physical: noise- and vibration-related illnesses.

ACTIVITY

Consider the accidents and ill-health associated with the industry/type of work with which you are involved.

Which are more likely to occur?
Which would potentially have the greatest severity of ill-health or injury?

IMMEDIATE AND UNDERLYING CAUSES OF AN ACCIDENT

The purpose of an accident investigation is to identify the immediate and underlying causes of an accident. In doing so, it will be possible to implement recommendations that can be developed to prevent the accident from happening again.

The four steps to investigate an accident are:

1. Gather factual information about the event.
2. Analyse the information and draw conclusions about the immediate and root causes.
3. Identify suitable control measures.
4. Plan the remedial actions.

Almost all accidents are caused by human error. 'The actions of people account for 96 per cent of all injuries' (DuPont).

Immediate causes of an accident are both the unsafe acts of individuals and unsafe workplace conditions. These include:

- Failure or breakdown of equipment or the use of incorrect tools.
- Failure to wear personal protective equipment and a poor standard of housekeeping.
- Failure to replace guards on machinery.
- Involvement of incompetent or unauthorized personnel.

Underlying (root) causes are referred to as failures in the management system or lack of management control conditions. These include:

- Failure to provide an acceptable level of training for operations where competence was required.
- A poor standard of supervision.
- Failure to complete risk assessments and introduce safe systems of work.
- Failure to recognize and manage the presence of stress in operatives arising from production issues.
- Inadequate procedures for routine maintenance operations.

ACTIVITY

Review an accident you have witnessed, been involved with or heard about. Then identify the immediate and underlying conditions.

RECORDING AND REPORTING

An organization should have a system in place for the internal reporting of accidents to enable the company to:

- Identify high-risk activities that require further control.
- Identify patterns of occurrence of accidents and incidents – types of machines, activity, part of the body injured, etc.
- Identify the immediate and underlying cause of the accident.

- Identify trends: either increasing or decreasing numbers of occurrence.
- Obtain information for civil claims.
- Provide statistics that help to form national health and safety policies.

Factors that may discourage employees from reporting accidents at work include:

- Ignorance of the reporting procedures.
- Lack of obvious management response to earlier reported accidents.
- Overcomplicated reporting procedures.
- Peer pressure.
- Possible retribution by management.
- To avoid receiving first aid or medical treatment (for whatever reason).
- To preserve the company's or department's safety record (particularly when an incentive scheme is in operation).

ACTIVITY

Review the policy for recording and reporting accidents in your workplace. Are they effective?

METHODS OF IMPROVING HEALTH AND SAFETY PERFORMANCE

Poor health and safety performance can be identified when there is evidence of:

- High accident rates.
- High level of absenteeism.
- Unsafe behaviour.
- High level of complaints from workforce.
- Poor housekeeping.

The above examples often occur as a result of underlying problems relating to a lack of resources, poor management commitment, lack of training and poor risk assessments. These become evident as unsafe behaviour and poor communication throughout the workplace.

The benefits of improving health and safety performance include:

- Monitoring and reviewing is a vital component of any safety management system and is essential if the system is externally accredited and audited by a specific body.
- To 'benchmark' the organization's performance against that of similar organizations or an industry norm.
- To assess compliance with legal requirements and accepted national and international standards.
- To be able to provide a Board of Directors or safety committee with relevant information.
- To boost morale and motivate the workforce.
- To compare actual performance with previously set targets.
- To identify any new or changed risks.
- To identify substandard health and safety practices and conditions.
- To identify trends in relation to different types of incident, or incidents in general, by analysis of relevant incident data.
- To identify whether control measures are in use, to assess their effectiveness and to be able to make decisions about appropriate remedial measures for any deficiencies identified.

ACTIVITY

Using this list of benefits identify the factors that have the most impact on your workplace's performance. Consider why this is the case.

RISK ASSESSMENT

The aims and objectives of a risk assessment are there to prevent accidents or ill-health occurring to workers or to others who may be affected by the work undertaken.

The risk assessment should be suitable and sufficient – which means that it:

- Should identify all significant hazards and risks arising from, or connected with, the activity to be carried out.
- Identify all the persons at risk, including employees, other workers and members of the public.
- Evaluate the adequacy and effectiveness of existing control measures.
- Identify other protective measures that may be required, enabling priorities to be set.
- Be appropriate to the nature of the work and be valid over a reasonable period of time.

The risk assessment should also be completed by someone who is competent (i.e. should have such practical and theoretical knowledge and actual experience of the machinery or plant he has to examine, so as to enable him to detect defects or weaknesses which it is the purpose of the examination to discover and to assess their importance in relation to the strength of the machinery or plant in relation to its function).

HSE outlines five steps to risk assessment:

Step 1	Identify the significant hazards.
Step 2	Decide who might be harmed and how.
Step 3	Evaluate risks and decide on precautions.
Step 4	Record your findings and implement them.
Step 5	Review the assessment and update if necessary.

Hazard

A source of energy with the potential to cause immediate injury to personnel and damage to equipment, environment or structure; or the properties of toxic substances, such as chemicals, gases or radioactivity, that may cause health problems immediately, or, in the short term or longer term, in people exposed to those substances.

Common types of safety hazards – with examples – include:

- Mechanical: moving parts of machinery.
- Trip: trailing cables.
- Slip: wet floors.
- Falls from height: insecure railings.
- Fire: hot work surfaces, damaged electrical cables.
- Electricity: electrical equipment.
- Manual handling: unsafe lifting techniques, heavy or unstable loads.
- Chemicals: explosive, corrosive substances.
- Vehicles: unsafe driving or work area.
- Work equipment: damaged or unsafe.
- Workplace: poor lighting.

Common health hazards – with examples – include:

- Ergonomic: muscular skeletal disorders.
- Physiological: stress, depression and/or anxiety.
- Respiratory: breathing or lung problems.
- Diseases: infections and skin.
- Physical: noise- and vibration-related illnesses.

Risk

The likelihood or probability that harm or loss will come about, taking into account the extent and severity of the outcome.

To calculate the risk of an identified hazard it is necessary to evaluate the likelihood and severity of the hazard. This can be calculated by multiplying the likelihood by the severity, as shown below.

Likelihood/Probability:

1 Unlikely: 1 in 10,000 chance of happening.
2 Likely: 1 in 1,000 chance of happening.
3 Very likely: 1 in 100 chance of happening.

Severity/Consequence:

1	Minor:	needing first aid.
2	Moderate:	needing ambulance.
3	Major:	leading to long-term injury/ill-health or death.

Likelihood × Severity = Level of risk; for example, if you evaluate that there is a **very likely** chance of an event occurring with a **moderate** severity the calculation would be 3 × 2 = 6. This is known as *quantitative risk assessment* i.e. using numbers to represent the level of risk. It is possible to use a five by five matrix, or even 10 by 10 for the likelihood and severity.

Another approach is *qualitative risk assessment*, which operates by allocating a high, medium and low rate to the likelihood and severity with the level of risk taking the highest rate. Therefore the above example a **very likely** chance of an event occurring with a **moderate** severity the calculation would be High × Medium therefore the risk level would be High; if the likelihood was Unlikely (low) and severity Major (high), or vice versa, this could be evaluated as a Medium Level of risk.

Once evaluated the level of risk the next step is to reduce the level risk so far as is reasonably practicable, this can be accomplished by reducing the likelihood and/or severity.

Control hierarchy

Control hierarchy is the ordering of available control measures based on the relative importance of each measure in protecting people. The hierarchy emphasizes the importance of engineering controls over administration, personal protective equipment and training.

There are various versions of the hierarchy of control – the underlying principle is that the hazard should be completely removed first (elimination); if this is not possible, perhaps the hazard can be changed for a less harmful hazard (substitution); physical barriers and ventilation systems should be considered next (technical controls); signage may be used to highlight any areas that may still pose a risk along with the development of safe systems of control (procedural controls) and finally, as a last resort, since it protects only the worker if used correctly, personal protective equipment (behavioural control).

1 Elimination.
2 Substitution.
3 Technical – engineering controls.
4 Procedural – signage/warnings and/or administration controls.
5 Behavioural – personal protective equipment.

Specific risk assessments that may be needed include:

- Computer – Display Screen Equipment (DSE).
- Hazardous substances, for example:
  - working with lead;
  - asbestos removal.
- Manual handling.
- Noise.
- Fire.
- Ergonomic – Assessment of Rapid Tasks (ART).
- Specific persons – young, pregnant, disabled.

ACTIVITY

Either review different types of risk assessments within your workplace or from the Internet, *or* undertake, with supervision, a risk assessment in an area about which you have some understanding/experience.

HEALTH SURVEILLANCE

The role of health surveillance is to systematically detect and assess the early signs of adverse effects on the health of workers exposed to certain health hazards.

Different methods of health surveillance are used depending on the type of work employees are undertaking.

Examples of types of work and suitable health surveillance are:

Confined spaces (use of respirators)	Occupational health medical and questionnaire.
Contact/use of substances hazardous to health	Varies depending on substance - may include: ■ Health questionnaire. ■ Respiratory function tests/spirometry tests. ■ Skin surveillance. ■ Blood tests. ■ Urine tests.
Display Screen Equipment	Eyesight test (every two years). DSE assessment.
Drivers	Driver assessment.
Ionizing radiations	Dosimetry and personal monitoring.
Laser users	Eyesight test.
Manual handling	Health questionnaire.
Night work	Occupational health assessment or questionnaire.
Noise	Audiometry assessment.
Pregnant workers	Occupational health assessment or questionnaire
Vibration	Vibration assessment.
Young workers	Health questionnaire.

ACTIVITY

Identify the health surveillance carried out at your workplace. Should any more be undertaken?

WORK EQUIPMENT

Scope of work equipment

Various hand tools may be used within the workplace. These may be non-powered or powered (e.g. chisel or electric drill). Each piece of equipment has its own individual hazards, in addition to hazards which the user brings (e.g. lack of training, inexperienced, etc.).

Access equipment includes ladders, working platforms (e.g. scissor lifts and mobile elevated working platforms). Specialist checks/inspections/ training may be required under national legislation for the use of working platforms.

Work equipment may be defined as 'any machinery, appliance, apparatus tools or assembly of components which are arranged so that they function as a whole'. Therefore equipment used in the workplace may be defined as work equipment.

Hazards relating to work equipment fall into two main categories:

1 Mechanical hazards.
2 Non-mechanical hazards.

Mechanical hazards

Mechanical hazards such as sharp blades, moving parts, etc. may result in the following injuries:

Type of injury	*Example*
Crushing	Scissor lift
Shearing	Sharp blade of a guillotine
Cutting or severing	Sharp blade of a saw
Entanglement	Drill bit turning
Drawing-in or trapping impact	Moving internal parts on a conveyor
Stab or puncture	High-power nail gun
Friction or abrasion	Conveyor belt
High-pressure fluid injection	Air hose from compressor

Non-mechanical hazards

Non-mechanical hazards such as electricity, generation of heat, noise or vibration may result in the following injuries:

- Electrocution/electrical burn/fire.
- Thermal burns.
- Noise-induced hearing damage.
- Vibration injuries.
- Radiation illnesses (ionizing and non-ionizing).
- Chemical injuries and ill-health (corrosive burns, dermatitis).

- Ergonomic ill-health – work-related upper limb disorder/injury to a worker's back.
- Slipping, tripping and falling.

Controls for work equipment

The hierarchy of control measures detailed above may be adapted to reduce the risk of work equipment causing injury; for example, elimination/substitution of work equipment, such as replacing a noisy drill with a quieter one.

Controls	*Example*
Technical controls/ engineering include:	■ Fixed guard. ■ Interlocking guard. ■ Self-acting guard (old name: automatic guard). ■ Adjustable guard. ■ Sensitive protective equipment (old name: trip device). ■ Two-handed controls. ■ Emergency stop.
Procedural controls include:	■ Information, instruction and training. ■ Restriction on use/maintenance. ■ Safety signs. ■ Warning alarms – audible and visual.
Behavioural controls include:	■ Personal protective equipment (PPE); for example, gloves, safety glasses. ■ Appropriate clothing; for example, overalls. ■ Hair tucked in when using work equipment. ■ Operators not being under the influence of drugs or alcohol. ■ Jewellery not worn when using work equipment.

All equipment provided for use at work should be:

- Accompanied by suitable safety measures.
- Inspected regularly.
- Maintained in a safe condition.
- Suitable for the intended use.
- Used only by people who have received adequate information, instruction and training.
- Equipment sold in the European Community must comply with various Directives and have the CE mark affixed to the equipment.

> **ACTIVITY**
>
> Identify the types of equipment at your place of work. What are the specific hazards associated with each, and what control measures have been implemented to combat the hazards?

VEHICLES IN THE WORKPLACE – TRANSPORT AND PEDESTRIANS

Hazards associated with the use of vehicles in the workplace may be categorized into four areas (some examples are included):

	Hazard
Vehicle:	■ Not maintained. ■ Unsuitable for the task. ■ Failure of breaks, lights, etc. ■ Overturning. ■ Collisions (with pedestrians, other vehicles or people).
Area/place of work:	■ Work area unsuitable; for example, poor road conditions. ■ Poor lighting. ■ Poor signage.
Operator/driver/ persons:	■ Pedestrians in area with vehicles. ■ Untrained operators. ■ Operators under the influence of drugs and/or alcohol.
Procedures:	■ Operators not keeping to speed limits/direction of traffic. ■ Operators not following pre-use checklists. ■ Pre-use checklists not produced.

Control measures

Elimination/substitution of the use of work vehicles should be considered (i.e. not having to use a vehicle in the first place or using public transport, for example, taking a train).

Technical controls/engineering include:

- Ensure vehicle is suitable for the task.
- Vehicle is fitted with reversing alarms, correct lighting, etc.
- Work area is suitable for the vehicles (enough space, traffic routes clearly identified).

- Segregation of pedestrians and vehicles.
- Limited change of surface (e.g. not having steep gradients).
- Safety signs displayed in the workplace/area.

Procedural controls include:

- Regular maintenance of vehicles by competent persons.
- Driver is suitably trained to operate the vehicle.
- Restriction on use/maintenance.
- Site rules developed, taught and enforced.

Behavioural controls include:

- Appropriate clothing provided and worn (e.g. overalls).
- Safety signs are followed (e.g. no speeding).
- Operators not to be under the influence of drugs or alcohol.
- Safety-belt to be worn/protection and restraint systems used.
- Pedestrians in area should wear hi-visibility clothing.

ACTIVITY

Review your workplace vehicle/transport policy. How could it be improved?

Complete a risk assessment or review your workplace vehicle/transport risk assessment.

The following not only illustrates the hazards, risks and controls for pedestrians in the workplace but the principles of carrying out a risk assessment.

Typical **hazards** found in the workplace that have an impact on pedestrians include:

- Wet floors, trailing cables, uneven carpets, leading to slips, trips and falls on the same level.
- Falls from a height.
- Collisions with moving vehicles.
- Being struck by moving, flying or falling objects.
- Striking against fixed or stationary objects.

Those at specific risk will include contractors, visitors, young persons, etc. Extra controls may be needed.

Risks associated with hazards found in the workplace that have an impact on pedestrians include:

- Moderate injury (broken bones, cuts and grazes) as a result of slips, trips and falls on the same level.
- Major injury/death as a result of a fall from a height.
- Major injury/death as a result of collision with a vehicle.
- Moderate injury (to the head or other body part) from being struck by moving, flying or falling objects.
- Minor injury (cuts and bruises) as a result of coming into contact with fixed or stationary objects.

Control measures for pedestrian hazards include:

- Slip-resistant surfaces/spillage control and drainage.
- Designated walkways.
- Fencing and guarding.
- Use of signs and personal protective equipment.
- Information, instruction, training and supervision – especially for those vulnerable persons identified above (contractors, visitors, young persons, etc.).
- Maintenance of a safe workplace.
- Cleaning and housekeeping requirements.
- Good/clear access and egress (exit).
- Environmental considerations (good lighting).

ACTIVITY

Review your pedestrian safety. How could it be improved?

Complete a risk assessment or review your pedestrian risk assessment.

ELECTRICITY

A damaged or faulty wired plug, damaged cable or equipment is an obvious hazard. However, electricity may be classed as a hazard in itself. Electricity can cause damage to equipment, fire or injury, and death. Secondary injury may be caused from falls from a height (e.g. if a worker has an electric shock while on a stepladder).

Possible effects of electricity running through the body range from a minor tingling sensation to more serious effects experienced after a severe electric shock:

- Cardiopulmonary effects, in particular, the risk of fatal injury due to disruption of heart rhythm.
- Interference with nerve/muscle action, leading to involuntary grip.
- Possibility of damage to internal organs.
- Tissue burns.
- Death.

Factors influencing severity include the individual (age, health conditions, if wet/if working outside, etc.) as well as voltage, frequency, duration, resistance and power.

Voltage

Voltage is the measure of the potential difference in electrical potential between the two terminals of a circuit and is measured in volts (V). It is the 'electrical pressure' that moves electrons through a conductor (i.e. wires, etc.). The range for domestic/office-based equipment for voltage is 220–240v in the UK, but other countries, such as the USA, use 110v current. Higher voltage systems are used by some heavy industry applications. Contact with 240v can be fatal.

Current

Current is the flow (movement) of an electric charge through a conducting body/rate of electrons moving past a point within one second and is measured in amperes (amps). The current may be direct current (DC) or alternating current (AC) – if direct then from a battery, if alternating from a mains supply. The frequency is how the current is transmitted from a power plant to the end user (50Hz in the UK). Depending on the equipment being used, the fuse will have a different amp rating (e.g. 5 or 13).

Resistance

Resistance is the force that reduces or stops the flow of electrons and opposes voltage, i.e. the degree to which an object opposes an electric current through it. Resistance is measured in ohms (Ω). Higher resistance will decrease the flow of electrons; lower resistance will allow more electrons to flow.

Power

In an electrical system power (P) is equal to the voltage multiplied by the current.

Power is measured in wattage (watts).

The standard hierarchy of control (elimination, substitution, technical, procedural and behavioural) may apply with electricity; however, it would be difficult to do most things if we eliminate its use. There are ways that it can be substituted in the workplace (e.g. using pneumatic equipment). However, this option has limited scope; for example, using candles or gas lamps creates more of a hazard than using an electric desk light!

However, electrical isolation, namely the disconnection and separation of electrical equipment from every source of electrical energy in such a way that this disconnection and separation is secure, is one of the best control measures that can be used; (e.g. during maintenance work).

Controls	*Example*
Technical controls/engineering include:	Equipment design: ■ Earthing. ■ Double insulation. ■ Reduced low voltage systems. Over current protection: ■ Fuse. ■ Circuit breaker. Electrical fault protection: ■ Residual current device. ■ Surge protection device.
Procedural controls include:	■ Testing and inspection. ■ Safe use (training).
Behavioural controls include:	■ Not leaving plugs switched on (mobile phone charger). ■ Not overloading sockets. ■ Not carrying out unauthorized repairs.

Earthing

Earthing is the bonding of metallic enclosures of electrical equipment, cable armouring, conduits and trunking, etc. so that these conductors are electrically continuous and securely connected to the general mass of earth at one or more points. Earthed conductors in a circuit provide a safe path for any fault current to be dissipated to earth. These will be connected to any exposed metal parts of a component connected to an electrical circuit that should not normally carry a current. Therefore, if a fault develops and becomes live, the earth conductor will carry the fault current away. Electrical equipment that is earthed is referred to as Class I.

Double insulation

Double insulation is a protective system where the live conductors of electrical equipment are covered by two discrete layers or components of insulation. This arrangement avoids the need for any external metalwork of the equipment to be connected to a protective conductor or to earth. Electrical equipment that is double insulated is referred to as Class II.

Reduced low voltage systems

Electrical supply systems in which the maximum voltage to earth that can occur in the event of a fault or damage to the system is reliably limited to a value which is unlikely to cause danger to people.

Fuse

The fuse is the thin wire, placed in a circuit, of such size as would melt at a predetermined value of current flow and therefore cut off the current to that circuit. It provides protection for the circuit and apparatus if excess current flows through it. Plugs are often fitted with a fuse (e.g. 5 or 13 amps).

Circuit breaker

A circuit breaker is an automatic device/switch used for stopping the flow of current in an electric circuit in the case of excess current flow. It is designed to protect the equipment from damage. It operates similar to a fuse but may be reset.

Residual current device (RCD)

A residual current device is an electro-mechanical device designed to interrupt the electrical supply to equipment if a fault is detected, hence minimizing injury in the event of a person receiving an electric shock.

Surge protection

This is a protection device that will regulate the flow of voltage to prevent damage to the equipment from voltage spikes/high voltage charges. This is often used on Information Technology equipment.

Inspection and testing of portable and transportable equipment

Electrical hazards that could be discovered by a visual inspection are as follows:

- A lack of circuit protection such as the absence of an appropriate fuse or a failure to use a residual current device.
- Appliances in a dirty or wet condition with vent holes filled with dust.
- Cuts, abrasions and cracks in inner and outer cable insulation.
- Damage to plugs and sockets and a failure to cord grip the cable.
- Damage to the outer case of an appliance and the absence of effective operating controls.
- Evidence of bare wires and conductors.
- Evidence of incorrect, unsafe or unauthorized repairs.
- No evidence of the provision of an adequate earth.
- Overloaded circuits and sockets.
- Portable appliance tests which are out of date.
- Incorrect choice of an appliance for the task to be carried out.

In the UK, Portable Appliance Testing (PAT) is common practice that involves regular testing of electrical work equipment. A competent person will check the plug for correct fuse, cable for damage and the equipment to see if it is working correctly. Once the visual check is completed the competent person will then check, using a PAT machine, the earth bond continuity (if earthed/Class I) and the insulation resistance.

Electrical equipment does not need to be portable appliance tested annually, or in some cases at all; for example, extra low voltage equipment (mobile phone chargers, etc.). A risk assessment must be undertaken to determine how likely the equipment is to fail in its use and how dangerous that effect would be.

General checks that can be made on both work equipment and in the workplace include:

- A lack of circuit protection (e.g. absence of an appropriate fuse or failure to use a residual current device).
- Appliances in a dirty or wet condition.
- Appliances with vent holes filled with dust.
- Cables not being uncoiled properly.

- Checking that outlets are not overloaded.
- Conforms with recognized standards (e.g. CE marking and European standards).
- Cuts, abrasions and cracks in inner and outer cable insulation.
- Damage to plugs and sockets.
- Damage to the outer case of an appliance.
- Evidence of bare wires and conductors.
- Evidence of incorrect, unsafe or unauthorized repairs.
- In vulnerable positions where equipment may suffer damage.
- Incorrect choice of an appliance for the task.
- Having an effective procedure for reporting defects or damage.
- Evidence of being subject to appropriate and regular fixed installation and portable appliance testing.
- Provided with means of isolation after use and that records are kept of maintenance that has been carried out.
- Need to ensure that all fuses are of the correct size.
- Evidence of the provision of adequate earthing.
- Overloaded circuits and sockets.

ACTIVITY

Review your use of electrical equipment within your own workplace. Identify how often portable appliance testing is carried out. Does this meet with current insurance and HSE guidelines?

Review chart may be found within http://www.hse.gov.uk/pubns/indg236.pdf.

FIRE

Consequences of fire can be devastating. They include:

- Building damage.
- Death.
- Environmental damage (e.g. from fire water, i.e. water used to put out the fire).
- Loss of business, production, jobs.

- Loss of flora and fauna (e.g. forest fires).
- Personal injury – burns, secondary injuries during escape.
- Structural failure of buildings.

Fire triangle

For a fire to start, it requires an ignition source, fuel and oxygen.

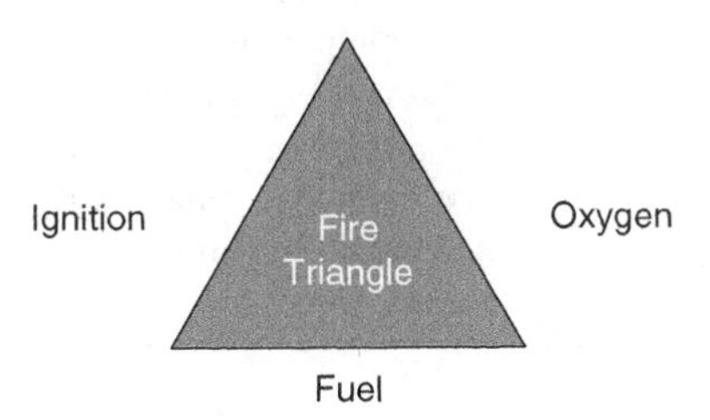

Ignition source

This is the starting point for the fire, i.e. what causes the fire. The following list identifies some of the common causes of fire in the workplace:

- Arson.
- Chemical reactions (exothermic).
- Electricity (faulty electrical equipment, sparks and static electricity).
- Heat sources (naked flames, hot surfaces, hot work, etc.).
- Smoking materials.

Oxygen sources

The main source of oxygen for fire is in the air around us (oxygen is approximately 20 to 21 per cent of the air we breathe). This may be via natural airflow through doors, windows and other openings, or through mechanical air-conditioning systems and air-handling systems.

Oxygen may also be found in materials used or stored in a workplace:

- Compressed air line(s).
- Ventilation systems.
- Oxidizing agents – nitrates, chlorates, chromates and peroxides.

Sources of fuel

These can be split into different classes: solids, gasses, metals and cooking oils.

Examples are:

Source	*Example*
Solids	■ Wood, paper, card. ■ Plastics, textiles. ■ Building materials.
Liquids	■ Flammable liquids – petrol, spirits, etc.
Gases	■ Flammable gases – liquid petroleum gas (LPG), acetylene, propane, butane, etc.
Metals	■ Combustible metals – magnesium, etc.
Cooking oils	■ Chip pan oil.

Classification of fires

Fires are classified according to the type of fuel that is burning. Should the wrong extinguisher be used on the wrong class of fire it may well make matters worse. Types of fires may be classified as follows:

Class	*Source*
Class A	**Solids**
Class B	**Liquids**
Class C	**Gases**
Class D	**Metals**
Class F	**Cooking oils**

It is worth noting that there is no class E. Fires that involve electrical equipment are not classified in the same way.

Transmission of heat/fire spread

Transmission of heat and fire spread are conduction, convection, radiation and direct burning.

Conduction

Conduction is the transfer of heat through solid materials involving the molecule-to-molecule transfer of heat through conducting solids such as metal beams or pipes to other parts of the building and igniting combustible or flammable materials.

Convection

Convection is the transfer of heat from a liquid or gas (i.e. air, flames or fire products) to a solid or liquid surface. Heat can be carried by rising air currents (convection), causing a buildup of hot gases.

Radiation

Radiation involves the emission of heat in the form of infrared radiation, which can raise temperatures of adjacent materials (e.g. electric fire elements).

Direct burning

Direct burning is when the flames/heat reach combustible materials and ignite them.

Reducing the risk of a fire

Various factors should be considered when trying to reduce the risk of a fire. These include:

- Control of smoking and smouldering materials.
- Good housekeeping to prevent accumulation of waste paper and other combustible materials.
- Regular lubrication of machinery to prevent sparks.
- Regular inspection of electrical equipment for damage.
- Ensuring that electrical systems are not overloaded.
- Isolating equipment that is not in use.
- Ensuring ventilation outlets on equipment are not obstructed.
- Controlling hot work by permits or by creation of designated areas.
- Providing proper storage facilities for flammable liquids away from sources of ignition.
- Segregating incompatible chemicals and implementing security procedures to reduce the risk of arson.

Safe storage and use of flammable liquids

Small quantities of dangerous goods may be found in most workplaces. Whatever they are used for, the storage and use of such goods can pose a serious hazard unless basic safety principles are followed. This can be accomplished by addressing the following five areas:

- Ventilation.
- Ignition.
- Containment.
- Exchange.
- Separation.

Ventilation

Is there plenty of fresh air where flammable liquids or gases are stored and used? Good ventilation will mean that any vapours given off from a spill, leak or release from any process will be rapidly dispersed.

Ignition

Have all obvious ignition sources been removed from storage and handling areas? Ignition sources may be varied, including sparks from electrical equipment or welding and cutting tools, hot surfaces, open flames from heating equipment, smoking materials, etc.

Containment

Are flammable substances kept in suitable containers? If there is a spill, will it be contained and prevented from spreading to other parts of the working area? Use of lidded containers and spillage catchment trays, for example, can help prevent spillages spreading.

Exchange

Can an exchange of a flammable substance for a less flammable one be made, i.e. substitution? Can flammable substances be eliminated from the process altogether? Are there other ways of carrying out the job more safely?

Separation

Are flammable substances stored and used well away from other processes and general storage areas? Can they be separated by a physical barrier, wall or partition? Separating hazards in this manner will contribute to a safer workplace.

Fire detection
Types of detectors that are commonly used in the workplace include:

- Thermal.
- Smoke.
- Flame.

The correct selection of fire detection devices will prevent false alarms. For example, it would not be appropriate to put a smoke detector in a kitchen – a heat or flame detector would be more appropriate.

Fire alarms
Fire alarms must be checked weekly to ensure that they are working. Manual call points should be sited on exit routes, 1.4m from floor level, conspicuous and not more than 30m apart. The alarm should be a minimum of 65dB (decibels) or 5dB above background noise likely to last for 30 seconds, and for sleeping 75dB at bed head height.

Evacuation/means of escape
Factors that need to be considered when evaluating means of escape include:

- An acceptable distance to the nearest available exit.
- Escape routes of sufficient width and fire protected.
- Clear signing of escape routes.
- Provision of emergency lighting.
- Escape routes kept clear of obstructions with fire doors closed to prevent the spread of smoke.
- Provision of fire-fighting equipment.
- Appointment and use of fire marshals/wardens.
- Procedures for evacuation of those with a physical impairment (in relation to hearing, sight or mobility).
- Identification of a safe assembly point.
- Practice of evacuation plan at regular intervals.

Fire doors
Fire doors should open in the direction of travel, be self-closing and be wide enough (e.g. for a wheelchair/number of persons likely to need to use the door in the event of an emergency). They should also be fire

resistant for at least 30 minutes. Fire doors should also be able to be opened at all times (i.e. not locked) to prevent exiting.

Emergency lighting

Factors to be considered when addressing emergency lighting include ensuring level of lighting meets national guidelines/legislation. In most cases emergency lighting is needed for:

- Corridor intersections.
- Exit doors.
- Change in floor level/over each stair flight or landing.
- Outside each final exit.
- Near to fire alarm call points, etc.

Checking emergency lighting includes:

- Daily: check all maintained lamps are lit and control panel displays normal command.
- Monthly: Test self-contained luminaires by simulation of power failure.

Fire marshals/wardens

A fire risk assessment will identify the number of fire marshals needed. Marshals will need to attend a specialist course.

Principal duties of the fire warden are to:

- Take appropriate and effective action if a fire occurs.
- Ensure escape routes are available for use.
- Identify fire hazards in the workplace.
- Record and report their observations.

If a fire is discovered, the fire warden should:

- Ensure alarm has been raised.
- Check that manufacturing processes have been made safe.
- Evacuate staff from building or area involved.

- Check that any staff members or visitors with disabilities are assisted as planned.
- Call reporting centre and give details of location, severity and cause of fire, if known.
- Fight fire if it is deemed safe to do so.

Emergency evacuation

Questions to consider when developing emergency evacuation procedures include:

- Will it be a full or phased evacuation?
- How often will drills be undertaken – once or twice per year?
- Are individual evacuation procedures needed – for disabled employees/visitors?
- How will checks be made during and after an evacuation?
- Who will provide/how will training be provided for employees and visitors?
- Who will inform/how will the fire authority be informed?

Fire extinction

Only attempt to extinguish a fire if it is deemed safe to do so, if the alarm has been raised and if the person has been trained to tackle a fire.

Fires can be extinguished by:

- Cooling.
- Smothering.
- Starving.

Cooling

Cooling is used to reduce the ignition temperature by taking heat out of the fire – using water to limit or reduce the temperature.

Smothering

Smothering will limit the oxygen available and prevent the mixture of oxygen and flammable vapour through the use of a foam extinguisher or a fire blanket.

Starving

Starving will limit fuel supply by removing the source of fuel, isolating the flow of flammable liquids or removing wood and textiles, etc.

Fire extinguishers

There is more than one kind of fire extinguisher to be used on different classes of fire and on electrical fires:

Water fire extinguishers

These are used for class A fires, not on electrical fires.

These are the most useful and cost-effective forms of fire-fighting equipment for general fires. As a standard rule there should be at least one water extinguisher on every floor of the workplace.

Water fire extinguishers are large, steel-bodied extinguishers with a capacity of nine litres. Although they are the most common form of fighting Class A fires, they may only be used on this class of fire and must never be used on any other kind of fire.

Foam fire extinguishers

These are used for both Class A and Class B fires, not on electrical fires.

These are usually for use on Class B fires, but may also be used on Class A fires. The layer of foam acts as a blanket, starving the fire of oxygen and allowing the burning liquid to cool, and also minimizes the risk of reignition. They should not be used on electrical fires.

Dry powder fire extinguishers

These are used for Class A, B, C and electrical fires.

They contain a fine white/blue powder and may be used on Class A, B, C and electrical fires. Be aware, however, that anyone who has a lung function disorder, such as asthma, may experience breathing problems when using these extinguishers. Dry powder fire extinguishers are very effective; however, they can cause damage to equipment in the vicinity of the fire and will render any food or drink products in the vicinity unsafe for consumption.

Carbon dioxide fire extinguishers

These are used for Class B and electrical fires.

They can be used on class B fires and on fires involving live electricity. They are ideal for use in kitchen areas because carbon dioxide does not damage or contaminate food, unlike dry powder fire extinguishers. The non-flammable gas takes away the oxygen element of the fire triangle, thereby suffocating the fire.

Wet chemical fire extinguishers

These are used for cooking oil fires.

They contain a wet chemical and can be used on cooking oils. Their contents cool and emulsify on contact with burning oil and seal the surface, preventing reignition, and are common in industrial kitchens. They should not be used on electrical fires.

A competent person should carry out the selection and positioning of extinguishers. Factors to consider are:

- Positioning on escape routes.
- In the vicinity of specific risks and suitable for the risks they protect.
- Positioning off the floor.
- Maximum 30 metres travel distance between each extinguisher.

In addition, extinguishers should be:

- Examined monthly.
- Annually inspected and tested.

The five steps to fire risk assessment are:

Step 1	Identify fire hazards.
Step 2	Identify people at risk.
Step 3	Evaluate, remove, reduce and protect from risk.
Step 4	Record, plan, inform, instruct and train.
Step 5	Review.

This may be expanded to:

<table>
<tr><td>Step 1</td><td>Identify fire hazards: involves looking at the workplace, having an understanding of work carried out there and identifying sources of:
■ Oxygen.
■ Ignition.
■ Fuel.

Special note should be made of:
■ Electrical sources of ignition.
■ Smoking arrangements.
■ Arson risk.
■ Portable heating appliances.
■ Cooking facilities.
■ Lighting (risk of ignition).
■ Housekeeping.
■ Contractors/others in the workplace.
■ Dangerous substances, i.e. having an explosive/flammable risk.</td></tr>
<tr><td>Step 2</td><td>Identify people at risk and review fire protection measures to ensure safety of:
■ Employees.
■ Non-employees.
■ Disabled/young persons, etc.

Consideration/evaluation of the following needs to be addressed:
■ Means of escape – number of exits/routes blocked?
■ How fire would spread/what will limit it spreading?
■ Lighting – is it suitable?
■ Fire signage – is it suitable?
■ Warnings/alarms – are they suitable?
■ Number and types of fire-extinguishing appliances used.
■ Automatic sprinkler system – is one needed?</td></tr>
<tr><td>Step 3</td><td>Evaluate, remove or reduce and protect from risk: this step involves addressing issues raised in Step 2 and ensuring delegation of responsibility for fire safety.

The following is also required:
■ Appointment of fire marshals/wardens.
■ Testing of alarm system (weekly).
■ Development of an action plan with outstanding issues that need to be dealt with.</td></tr>
<tr><td>Step 4</td><td>Record, plan, inform, instruct and train:
Details of risk assessment should be available for all employees/others in the workplace. This needs to be supported by carrying out drills and fire training suitable to the workplace and its activities

Records need to be kept of:
■ Fire drills.
■ Fire training.
■ Alarm testing.
■ Emergency lighting.</td></tr>
</table>

Step 5	Review risk assessment. A review should be carried out annually or if there is a change in activity, or premises, that will have rendered the assessment invalid, or in the event of a fire!

ACTIVITY

Review the fire risk assessment at your workplace based on the above information. Then request the fire risk assessment template from the local fire brigade authority in your area (e.g. Cleveland Fire Brigade), and download their documents and complete them.

ERGONOMIC FACTORS: MANUAL HANDLING

Manual handling is the transportation or support of a load (including lifting, putting down, pushing, pulling, carrying or moving) by hand or through bodily force.

Common types of manual handling injuries include:

- Muscular sprains and strains.
- Back injuries:
  - compressed disc;
  - prolapsed/slipped disc.
- Trapped nerves.
- Hernia.
- Cuts, bruises and abrasions.
- Fractures.
- Work-related upper limb disorders (WRULDs).

Manual handling injuries may be classed as acute or chronic.

Acute

Acute injury occurs after limited exposure and shortly (hours, days) after exposure, such as someone picking up a box that causes immediate back pain, i.e. a traumatic injury.

Cumulative/chronic

Chronic injury occurs after prolonged exposure, such as carrying out a repetitive task.

Factors that increase the risk of manual handling injuries include:

- Lifting heavy or awkward loads.
- Applying excessive force.
- Awkward or fixed postures for work.
- Repetitive movements.
- Duration – extended time.
- Working environment – poor lighting.
- Other factors, including individual medical conditions.

An employer should:

- **Avoid** the need for hazardous manual handling, so far as is reasonably practicable.
- **Assess** the risk of injury from any hazardous manual handling that cannot be avoided.
- **Reduce** the risk of injury from hazardous manual handling, so far as is reasonably practicable.

Manual handling risk assessment

Factors to consider when carrying out a manual handling risk assessment include looking at the task, individual, load and environment (TILE) and asking the following questions:

Task – does it involve:	■ Holding loads away from trunk? ■ Twisting? ■ Stooping? ■ Reaching upward? ■ Large vertical movements? ■ Long carrying distances? ■ Strenuous pushing or pulling? ■ Unpredictable movement of loads? ■ Repetitive handling? ■ Insufficient rest or recovery? ■ A work rate imposed by a process?

Individual – does the job:	■ Require unusual capability? ■ Hazard to those with a health problem? ■ Hazard to those who are pregnant? ■ Call for special information/training? ■ Does the individual's clothing hinder movement or posture? ■ Does the individual's personal protective equipment hinder movement or posture?
Loads – are they:	■ Heavy? ■ Bulky/unwieldy? ■ Difficult to grasp? ■ Unstable/unpredictable? ■ Intrinsically harmful (e.g. sharp/ hot)?
Environment – are there:	■ Constraints on posture? ■ Poor floors? ■ Variations in level? ■ Hot/cold/humid conditions? ■ Strong air movements? ■ Poor lighting conditions?

Control measures that should be implemented using the TILE principle include the following:

Task

- Avoid tasks which require twisting the body wherever possible.
- Avoid bending and stooping to lift a load, as this significantly increases the risk of a back injury.
- Carrying distances should be minimized, especially if the task is regularly repeated.
- Items should ideally be lifted from no lower than knee height to no higher than shoulder height as outside this range lifting capacity is reduced and the risk of injury is increased.
- Items that are pushed or pulled should be as near to waist level as possible. Pushing is preferred, particularly where the back can rest against a fixed object to give leverage.
- Repetitive tasks should be avoided whenever possible.
- Tasks which involve lifting and carrying should be designed in such a way as to allow for significant rest breaks (rotation of tasks) to avoid fatigue.
- When items need to be lifted from above shoulder height a stand or suitable means of access should be used.

Individual

- Consideration should be given to age, body weight and physical fitness.
- Regard should be given to personal limitation. Employees should not attempt to handle loads that are beyond their individual capability. Assistance must be sought where this is necessary.
- Persons with genuine physical or clinical reasons for avoiding lifting should be made allowance for, as should pregnant women, who should not be required to undertake hazardous lifting or carrying tasks.
- Significant knowledge and understanding of the work is an important factor in reducing the risk of injury.
- Individuals undertaking lifting or carrying should be given suitable instruction, training and information to undertake the task with minimum risk.

Load

- An indication of the weight of the load and the centre of gravity should be provided where appropriate.
- Ensure there are secure handholds.
- Load should be kept as near as possible to the body to reduce strain and should not be of such size as to obscure vision.
- Unstable loads should be handled with particular caution as the change in the centre of gravity is likely to result in overbalancing.
- Use gloves where necessary to protect against sharp edges or splinters.

Environment

- Adequate space should be provided to enable the activity to be conducted safely, and the transportation route should be free from obstruction.
- Lighting, heating and weather conditions must be taken into account.
- Floors and other working surfaces must be in a safe condition.
- Adequate ventilation is required, particularly where there is no natural ventilation.

- Use of personal protective equipment may be necessary while carrying out manual handling activities. If the use of personal protective equipment restricts safe and easy movement, this should be reported.

Good handling techniques for lifting/principles of lifting are:

1. Think before you lift.
2. Keep the load close to your waist.
3. Adopt a stable position.
4. Ensure a good hold on the load.
5. At the start of the lift, moderate flexion (slight bending) of the back, hips and knees is preferable to fully flexing the back (stooping) or the hips and knees (squatting).
6. Don't flex your spine any further as you lift.
7. Avoid twisting the trunk or leaning sideways, especially while the back is bent.
8. Keep your head up when handling.
9. Move smoothly.
10. Don't lift more than you can easily manage.
11. Put the load down, then adjust.

ACTIVITY

Consider the activities at work that involve manual handling – are there safer ways to carry them out?

KEY NOTE

When you read the section about answering exam questions, you will see some repetition of the above information within the section on manual handling. As you read this information again you should start to remember some of the above!

ERGONOMIC FACTORS: REPETITIVE PHYSICAL ACTIVITIES/ RAPID TASKS

Repetitive tasks are made up of a sequence of upper limb actions of fairly short duration, which are repeated over and over again and are almost always the same (e.g. stitching a piece of cloth, manufacturing one part, packaging one item).

Factors to consider when carrying out an assessment of repetitive tasks include looking at the following areas and asking the following questions:

Shoulder arm movements	■ Does the task involve frequent or very frequent shoulder and arm movements (e.g. regular movement with some pauses or almost continuous movement)?
Repetition	■ Does the task involve similar motion patterns being repeated frequently?
Force	■ Does the task require moderate or strong force to be exerted?
Head/neck posture	■ Does the task involve holding the neck bent or twisted for part of the time or more than half of the time?
Back posture	■ Does the task involve the body being bent forward, sideways or twisted for part of the time or more than half of the time?
Shoulder arm posture	■ Is one or both of the elbows raised away from the body for part of the time or more than half the time?
Static shoulder and elbows	■ Is one or both of the shoulders and elbows in a static position (i.e. infrequently moved) for more than an hour?
Wrist posture	■ Is one or both of the wrists bent or deviated for part of the time or more than half of the time?
Hand and finger grip	■ Is a pinch or wide finger grip being used for part of the time or more than half of the time?
Static fingers, hand and wrist	■ Are one or both hands and wrists in a static position (i.e. infrequently moved) for more than an hour?
Breaks	■ Does the worker conduct the task continuously for more than an hour?
Work pace	■ Is it sometimes or often difficult to keep up with the work?

ACTIVITY

Consider the activities at work that involve repetitive tasks – are there safer ways to carry them out?

ERGONOMIC FACTORS: DISPLAY SCREEN EQUIPMENT

Symptoms that develop from computer use include: upper limb disorders (ULDs) affecting the arms, from the fingers to the shoulder and neck, visual problems and increased levels of stress.

Factors to consider when carrying out a DSE assessment include looking at the following areas and asking the following questions:

Display screens	■ Is the monitor in good working condition, including swivel and tilt functions and screen quality? ■ Is the screen size suitable for all tasks and are characters easy to read from the working distance? ■ Is the screen free from distracting reflections or glare? ■ Does the user know how to adjust brightness and contrast? ■ If there is a multi-screen set-up is it well configured for efficiency and safe working postures?
Keyboard	■ Is the keyboard separate from the screen? ■ Is the keyboard in good working condition, including tilt, and are the keys easy to read? ■ Is the keyboard comfortable to use?
Mouse	■ Is the mouse separate from the screen? ■ Is the mouse in good working condition? ■ Is the mouse comfortable to use, including pointer speed?
Software	■ Does the software work reliably? ■ Is the software suitable for the tasks?
Document work	■ Does the user have a suitable document holder, if needed?
Desk	■ Is the work surface free from clutter and well organized? ■ Is the work surface large enough for all items to be correctly positioned? ■ Is there sufficient off-desk storage for files and folders? ■ Is the area under the desk free from obstructions?
Chair	■ Does the chair have a working seat height adjustment? ■ Does the chair have a working backrest height adjustment? ■ Does the chair have a working backrest tilt adjustment? ■ Does the chair have a working swivel mechanism? ■ Does the chair have a stable five-star base with suitable castors?
Working posture	■ Is the user's chair comfortable, and is the chair in a supportive, upright position with good lumbar support? ■ Can the user sit at a safe working height and distance relative to the desk with feet fully supported? ■ Is the DSE set up appropriately for the user to adopt safe working postures? ■ Can all other items such as the telephone be used and accessed safely, including a headset if required?

Laptop	■ If a laptop is in regular, prolonged use at work does the user have a safe set-up? ■ If a laptop is in regular use at home does the user have a safe set-up? ■ If a laptop must be regularly carried, does the user have a suitable carrying case?
Environment	■ Is the lighting suitable and sufficient, without visual concerns, including adjustable window coverings fitted where necessary? ■ Confirm that the user is not affected by cold draughts and that air temperature is mostly acceptable. ■ Confirm that there is no distracting noise caused by equipment. ■ Can the user safely enter and exit the work station?
Individual	■ Does the user know about free eye test policy?

ACTIVITY

If your work involves using a computer, undertake a basic DSE assessment by asking yourself the above questions. What further action is needed?

HAZARDOUS SUBSTANCES

Common hazardous substance

Types of common hazardous substance and their ill-health effects include the following:

Wood dust
Inhalation of the dust arising from sanding or cutting operations may cause allergic or non-allergic respiratory symptoms and cancer.

Silica
Inhalation of silica may result in silicosis, which is a fibrosis of the lung.

Cement dust
Prolonged or repeated exposure to cement dust can lead to silicosis. Continuous contact with wet cement may cause burns or skin ulcers.

Solvents
Inhalation of vapours from organic solvents such as trichloroethylene being used as a cleaning agent can cause drowsiness and can depress the central nervous system, and may lead to liver failure.

Lead
Lead poisoning results from the inhalation of fumes produced from the heating of lead such as the oxyacetylene cutting of metal coated with lead paint.

Forms of substances

Hazardous substances can come in many different forms (i.e. their shape), as outlined below:

Liquids	Are in liquid form at normal room temperature and atmospheric pressure.
Gases	Are formless chemicals which occupy the area in which they are enclosed.
Vapors	Are a gaseous form of a liquid below its boiling point.
Mists	Are finely suspended droplets.
Aerosol	Are fine suspensions of solid particles or droplets in a carrier gas.
Fumes	Are fine particulate solids.
Dusts	Are solid particles of varying size.

ACTIVITY

Identify examples of hazardous substances that take the form of:

- Liquids.
- Gases.
- Vapours.
- Mists.
- Aerosols.
- Fumes.
- Dust.

Classification of substances

Substances having dangerous health effects are classified under one or more of the following categories:

Very toxic	Substances and preparations which, in very low quantities, cause death or acute/chronic damage to health when inhaled, swallowed or absorbed via the skin.
Toxic	Substances and preparations which, in low quantities, cause death or acute/chronic damage to health when inhaled, swallowed or absorbed via the skin.
Harmful	Substances and preparations which may cause death or acute/chronic damage to health when inhaled, swallowed or absorbed via the skin.
Corrosive	Substances and preparations which may, on contact with living tissues, destroy them.
Irritant	Non-corrosive substances and preparations which, through immediate, prolonged or repeated contact with the skin or mucous membrane, may cause inflammation.
Sensitizing	Substances and preparations which, if they are inhaled or if they penetrate the skin, are capable of eliciting a reaction by hyper-sensitization such that on further exposure to the substance or preparation, characteristic adverse effects are produced.
Carcinogenic	Substances and preparations which, if they are inhaled or ingested or if they penetrate the skin, may induce cancer or increase its incidence.
Mutagenic	Substances and preparations which, if they are inhaled or ingested or if they penetrate the skin, may induce heritable genetic defects or increase their incidence.
Toxic for reproduction	Substances and preparations which, if they are inhaled or ingested or if they penetrate the skin, may produce or increase the incidence of non-heritable adverse effects in the progeny and/or an impairment of male or female reproductive functions or capacity.

ACTIVITY

List the hazardous substances that are used in your workplace, showing the:

Substance / Form / Classification

ROUTES OF ENTRY OF HAZARDOUS SUBSTANCES INTO THE BODY

Absorption – through the skin or eyes

Skin contact is often the unrecognized route of entry into the body. While it provides reasonable protection against particulate matter, it provides very little protection of entry for substances such as solvents.

Inhalation – breathing in

Consideration must be given to the body's natural defence mechanism against inhaled particles. The ciliary escalator carries foreign bodies up the respiratory tract where they are either spat out or swallowed.

Ingestion – eating and drinking

An estimated 50 per cent of the particles deposited in the upper respiratory tract and 12.5 per cent from the lower passages are eventually swallowed.

Injection – puncturing skin; for example, needle

Needle injuries are common for health care workers, police and other public workers.

ACTIVITY

What is the most common form of entry for hazardous substances in your workplace, and why?

Biological agents

Biological agents are micro-organisms (fungi, bacteria and viruses) which may cause infection, allergy or toxicity, or otherwise create a hazard to human health.

ACTIVITY

List any biological hazards that you may be at risk from, due to the nature of your work.

Identify examples of biological agents:

- Fungi.
- Bacteria.
- Virus.

Factors to consider when assessing hazardous agents:

- Potential route(s) of entry – how the substances can enter the body (e.g. inhalation, skin absorption).
- Effects they may cause (e.g. dermatitis).
- Physical properties of hazardous agents.
- Quantities of hazardous agents used.
- Ways in which the solvent is applied (by hand/brush/spraying?).
- Frequency and duration of exposure to hazardous agents.
- Work methods employed.
- Who is being affected – the number of individuals exposed, their health status, pregnancy, etc.
- Information – manufacturers' material safety data sheets, guidance documents, etc.
- Potential additive/synergistic effects from other hazardous agents.
- Existing control measures in terms of their nature and adequacy.
- Identify the best form of atmospheric monitoring (AM) or biological monitoring (BM) for the hazardous agent.

Atmospheric monitoring (AM) involves the monitoring of airborne workplace contaminants through personal monitoring or, on occasion, through static monitoring. AM has limitations in that it excludes skin and ingestion as routes of absorption. AM may take a variety of forms with a number of available techniques; it is generally non-invasive.

Biological monitoring (BM) involves the measurement of indicators of the uptake of toxic substances in the body, such as in blood, urine or exhaled breath, in order to prevent health impairment. BM includes non-occupational exposure and takes into account individual dose and metabolism. BM is useful where there is a high dependency on personal protective equipment (PPE).

After gathering data about the hazardous substance (i.e. from the safety data sheets) a risk assessment for hazardous substances needs to be undertaken. This will include completing a pro-forma assessment with headings, such as:

- Date or internal company reference.
- Company information about product and/or trade name.
- Properties (i.e. vapour pressure, dustiness, etc.).
- Composition (i.e. proportion of the different components).
- Hazard and/or risk classification.
- Any short- or long-term exposure limits (for the country in which it is to be used).
- Description of the task(s).
- Details of the exposure profile.
- Details of the number of people exposed.
- Outline of the control measures being used.
- Details of the handling and storage facilities and processes.
- Details of the disposal of excess and waste material.
- Precautions and emergency procedures.
- Assessment of residual risk after control – high/medium/low.
- Results of airborne monitoring.
- Any further action(s) required.
- Review date (one year or less).

Once information has been gathered, control measures can be designed and implemented. These will be included in the risk assessment and if further action is needed this too must be recorded.

Hierarchy of control for hazardous substances

The following outlines the hierarchy of control for using substances hazardous to health. The following questions need to be addressed:

Eliminate
Can the hazardous substance be eliminated?

For example:

- Pigeon droppings are a hazardous substance – if you stop pigeons roosting (e.g. by using netting), you will prevent the buildup of droppings, meaning that people will not be exposed to them or need to use potentially hazardous products when cleaning up the droppings.
- Lead has been removed from petrol in many countries due to its hazardous properties.
- Use water-based paints so that you don't need to use white spirit or thinners.

Substitute

Can a less hazardous substance be substituted for a particular task?

For example:

- Put down an insecticide gel rather than using powder or spray, which may drift in the air and be inhaled or ingested by young children, who may mistake it for sherbet when it is used in nurseries, etc.
- Use a lower concentration of the hazardous substance – such as Milton fluid to disinfect surfaces rather than bleach (same chemical, just a lower concentration).
- Use chlorine tablets to disinfect swimming pools rather than powder, reducing the risk of inhalation of dust when handling the chemical.
- Use water-based paints rather than solvent-based paints (with the added benefit that thinners are not needed to clean brushes).

Technical engineering controls

Can the development of technical engineering controls reduce the risk?

For example:

- Capturing emissions by using dust bags on sanders and planers or HEPA filters on vacuum cleaners (remember to also consider how you will handle and dispose of collected emissions safely).
- Distance yourself from the substance; for example, by using a long-handled wiping tool in printing rather than applying the substance to a rag in your hand.
- Enclosing the substance by putting lids back on containers when not in use.

- Enclosing the work process by working with the substance in a fume cupboard or glove box.
- Extracting emissions using Local Exhaust Ventilation (LEV), such as an extractor hood on a circular saw to remove sawdust as it is produced.
- Where release cannot be contained (e.g. when painting), use general ventilation by opening doors and windows (non-toxic emissions only).

Procedural controls

Is it possible to change the work process or system to emit less of the substance or limit associated dangers?

For example:

- Brushing on paint rather than spraying it.
- Using water to suppress dust.
- Closing windows and/or doors to prevent air currents from blowing substances around.
- Storing less of the substance on the premises.
- Storing and using substances in different locations to prevent them from mixing with substances that could react dangerously if mixed together; for example, before using bleach to clean a toilet, flush the toilet to remove any other substance that may be in the bowl.

Behavioural controls

Workers must take responsibility for themselves.

For example:

- Following the information, instruction and training provided.
- Using all protective measures including wearing personal protective equipment (PPE).

PPE should always be the last resort as it only protects the person wearing it, and it is the most likely control to fail in its protection.

Main reasons for the failure of PPE include:

- Can be heavy or cumbersome, restricting movement and causing wearer to become hot.
- Inadequate fit for wearer.
- Incompatible with other PPE or personal items (e.g. goggles worn by workers with glasses).
- Interferes with other senses (e.g. sense of smell when wearing a mask).
- Must be used correctly – may require special training.
- Pressure to get the work done – it may be perceived that it takes 'too long' to put on all PPE needed.
- Requires maintenance and storage.

Hazardous substance risk assessment

Factors to consider when carrying out a hazardous substance risk assessment include looking at the substance, individual, task and environment (SITE) and asking the following questions:

Substance	■ Are there any exposure limits? ■ Could the substance be mixed (intentionally/accidentally) with another, which could be dangerous? ■ How often may people be exposed? ■ How will it get into the body? ■ Might the substance produce dangerous fumes or be a fire risk? ■ In what form is the substance (solid, liquid, airborne)? Would this lead to significant exposure? ■ What harm could it do to the body? ■ What is the chance of exposure? This may be affected by the training and information people have and how reliable and suitable the control measures are.
Individual	■ Are there any people who would be particularly vulnerable? ■ Are users capable of using the substance safely? ■ Who will be using/be exposed to the substance?
Task	■ How will the substance be used or produced? ■ What other hazards may be involved? ■ What will the user do next? ■ Will a substance be used in line with the manufacturer's instructions?

Environment	■ Is there anything in the environment that could be incompatible with the substance (e.g. naked flames in the presence of flammable substances or other substances that could chemically react with it)? ■ Where is the substance going to be used/produced and who might it affect?

ACTIVITY

Review a COSHH assessment from your workplace using the above questions.

WELFARE AND THE WORK ENVIRONMENT

Basic requirements at work with regard to welfare and the work environment include providing adequate welfare facilities and ensuring that the workplace/environment is safe. This will include having good levels of housekeeping.

Provision of adequate welfare arrangements is important both in terms of complying with national legislation and improving the morale of workers.

Factors to consider regarding welfare and the work environment include:

- **Maintenance of workplace, equipment, devices and systems** – including ventilation systems, emergency lighting, fencing, power doors, etc.
- **Ventilation** – provision of sufficient quantity of fresh or purified air. An audible fault warning alarm to be fitted to any equipment for this purpose.
- **Temperature in indoor workplaces** – reasonable temperature with sufficient thermometers around, except where very high or low temperatures are necessary for the job (e.g. in a cold store).
- **Lighting** – to be suitable and sufficient, preferably natural. Emergency lighting to be installed where risks may be created by light failure.
- **Cleanliness and waste materials** – walls, floors and ceilings to be capable of being kept clean. Wastes not to accumulate, but kept in correctly assigned bins, etc.

- **Room dimensions and space** – each person to have sufficient floor area, height and unoccupied space, usually around 11 cubic metres, except for kiosks, etc.
- **Workstations and seating** – suitable workstations and seating with adequate weather protection and provision for rapid escape in an emergency.
- **Condition of floors and traffic routes** – suitably constructed, not slippery or uneven and adequately drained. Barriers for holes, edges and slopes.
- **Falls and falling objects** – to prevent falls and falling objects.
- **Windows, transparent and translucent doors, gates and walls** – to be protected against breakage and marked to make visible. Safety glazing.
- **Windows, skylights and ventilators** – to provide safe opening methods and not create risk. Barrier windows lower than 800mm above floor.
- **Ability to clean windows** – to provide safe cleaning methods.
- **Organization, etc. of traffic routes** – to ensure pedestrian safety by separation and properly identified traffic routes. Protection of pedestrians from reversing vehicles.
- **Doors and gates** – to be of suitable construction and designed to avoid injury whether sliding, upward opening or powered. Vision panel in two-way doors.
- **Escalators and moving walkways** – equipped with suitable safety devices and emergency stops.
- **Sanitary conveniences** – sufficient numbers at readily accessible locations, separate male/female facilities, adequately lit, ventilated and kept clean.
- **Washing facilities** – sufficient and nearby conveniences, with hot and cold running water. Soap, towels or dryers to be provided. Separate male/female facilities.
- **Drinking water** – an adequate supply to be available and, if not obvious, should be labelled as 'drinking water'.
- **Accommodation for clothing** – dirty and clean clothing storage facilities. Drying facilities should also be provided.
- **Facilities for changing clothes** – if special clothing used, separate male/female changing facilities are required.
- **Facilities for rest and to eat meals** – suitable and sufficient facilities to enable persons to eat away from workplace where there may be risk of contamination.

ACTIVITY

Consider your place of work – are all the above factors taken into account?

STRESS

Occupational stress may be defined as an adverse reaction brought about by the responsibilities associated with work activities, or caused by conditions that are based in the corporate culture, or personality conflicts within the workplace. As with other forms of tension, occupational stress can eventually affect an individual physically, emotionally or behaviourally. As stress begins to take its toll, a variety of symptoms may result. These include:

Physical	*Emotional*	*Behavioural*
▪ Breathlessness. ▪ Headaches. ▪ Fainting spells. ▪ Chest pains. ▪ Tendency to sweat. ▪ Nervous twitches. ▪ Cramps or muscle spasms. ▪ Pins and needles. ▪ High blood pressure. ▪ Feeling sick or dizzy. ▪ Constant tiredness. ▪ Restlessness. ▪ Sleeping problems. ▪ Constipation or diarrhoea. ▪ Craving for food. ▪ Indigestion or heartburn. ▪ Lack of appetite. ▪ Sexual difficulties.	▪ Aggressive. ▪ Irritable. ▪ Depressed. ▪ Feeling bad or ugly. ▪ Fearing diseases. ▪ Fearing failure. ▪ Dreading the future. ▪ A loss of interest in others. ▪ Taking no interest in life. ▪ Feeling neglected. ▪ Feeling that there is no one to confide in. ▪ A loss of sense of humour.	▪ Have difficulty making decisions. ▪ Avoiding difficult situations. ▪ Frequently crying. ▪ Have difficulty concentrating. ▪ Biting nails. ▪ Denying there's a problem. ▪ Unable to show true feelings.

Sources of information that may be used to assess the risks of stress within the workplace include:

- Discussions at safety committees or team briefings.
- Feedback from face-to-face discussions between staff and line managers or supervisors, either informally or formally, for example, at performance appraisal reviews.
- Feedback from structured staff questionnaire with the appropriate analysis.
- Formal or informal complaints.
- Published guidance.
- Sickness and/or absence data and return-to-work interviews.
- Staff turnover data and exit interviews.

Management and control of stress at work – factors that cause stress

The organizational factors that can contribute to work-related stress are addressed in HSE Management Standards for Stress as follows:

Standards	*Example of stressors*
Demands	■ Excessive monitoring of work performance/rate (e.g. call centres). ■ Excessive workloads being placed on individuals. ■ Extremely demanding deadlines and targets. ■ Long hours/shiftwork causing disruption to daily routines. ■ Working environment – deficiencies in the working environment such as poor housekeeping, dirty conditions, extremes of temperature, noise, poor lighting, etc.
Control	■ A lack of control over task. ■ A lack of control over work scheduling. ■ Employee has no say in the way work is done/no consultation. ■ Forced pace of work. ■ Lack of policies. ■ No allowance for domestic issues (e.g. children's dental appointments, etc.) ■ No flexibility in start/finish times/breaks. ■ No possibility of changing tasks/variation(e.g. production line workers).
Support	■ Employees do not know where to go for support. ■ Lack of information/training given to employees. ■ Lack of management and peer support. ■ No response to requests for help and support. ■ Poor organizational culture.

Relationships	■ Aggressive/autocratic management style. ■ Atmosphere of conflict. ■ Customer/client pressures (e.g. complaints, threats of violence). ■ Harassment/discrimination/bullying issues. ■ Level of supervision (oppressive/insufficient/no support). ■ Poor relationship with supervisors/peers.
Role	■ Lack of clarity about objectives and individuals' roles and responsibilities. ■ Poor leadership.
Change	■ Conflict between individuals. ■ Constant restructuring/reorganization/change/uncertainty. ■ Fear of redundancy. ■ Lack of understanding of job requirements.

Management and control of stress at work – control measures

Using the above table, it is possible to identify how management can prevent and reduce the risk of stress in the workplace:

Standards	*Controls*
Demands	■ Reduce monitoring of work performance/rate (e.g. call centres). ■ Reduce workloads being placed on individuals. ■ Set realistic deadlines and targets. ■ Develop work–life balance. ■ Improve working environment – good housekeeping, air conditioning, reduce noise, improve lighting, etc.
Control	■ Improve control over task. ■ Improve control over work scheduling. ■ Allow employee to have a say in the way work is done/involved in consultations. ■ Review forced pace of work tasks. ■ Develop policies. ■ Provide for domestic issues (e.g. children's dental appointments, etc.). ■ Develop flexibility in start/finish times/breaks. ■ Discuss possibility of changing tasks/variation (e.g. production line workers).
Support	■ Employees given support person (i.e. line manager). ■ Provide information/training for employees. ■ Provision of management and peer support. ■ Ensure response to requests for help and support is provided. ■ Improve organizational culture.

Relationships	■ Improve management style. ■ Remove atmosphere of conflict. ■ Develop customer/client policies. ■ Develop harassment/discrimination/bullying policies. ■ Improve levels of supervision/offer support. ■ Improve relationship with supervisors/peers.
Role	■ Provide clarity about objectives and individuals' roles and responsibilities. ■ Improved positive leadership.
Change	■ Provide mediation between individuals if conflict arises. ■ Reduce levels of restructuring/reorganization/change/uncertainty. ■ Keep employees informed of any redundancy plans. ■ Provide job specifications.

Stress risk assessment

This will include a statement of the foreseeable situations that may arise whereby workplace stress poses a significant health risk and recognize the legal duties of the employer. It will outline the risks so far as is reasonably practicable.

The **five steps** to risk assessment may be adapted to produce a specific risk assessment for stress management:

Step 1	**Identify the significant hazards**	Consider activities that are being undertaken at the workplace that might pose a risk of occupational stress to employees (i.e. demand; control; support; relationships; role and change).
Step 2	**Decide who might be harmed and how**	Identify those exposed to workplace stressors and how they may be affected.
Step 3	**Evaluate risks and decide on precautions**	Evaluate each specific risk – demand; control; support; relationships; role and change – and identify specific and general controls.
Step 4	**Record and implement your findings**	Record any significant findings of the assessment and implement them.
Step 5	**Review the assessment and update if necessary**	Review and revise the assessment at regular intervals or more frequently if required.

ACTIVITY

Review your workplace stress. How could it be improved?

Complete a risk assessment or review your workplace stress risk assessment.

VIOLENCE

Workers who are more at risk of violence in the workplace include those who handle money (e.g. cashiers, shop assistants, etc.), and those who deal with the public (e.g. emergency services, shop assistants, etc.). Employers must ensure the safety of their workers, protecting them from the risk of violence.

Prevention strategies that an organization could consider to reduce the risk of violence towards workers include:

- Introducing procedures for the reporting of incidents.
- Job designed with a reduction in cash holdings and introduction of appointment systems and team working where practicable.
- Procedures for lone and out-of-hours working and for home and off-site visits.
- Providing employees with training in confrontation management and stress reduction techniques.
- Secure areas with coded locks and physical barriers, and the use of check-in and check-out procedures.
- Completion of initial risk assessments by a competent person.
- Design of workplace and public areas to restrict access to non-authorized persons.
- Use of trained security staff.
- Use of security equipment such as CCTV, alarm systems, personal alarms and panic buttons.

> **ACTIVITY**
>
> Review your workplace violence policy. How could it be improved?
>
> Complete a risk assessment or review your workplace violence risk assessment.

NOISE

The ear

- **Outer ear** collects sound waves and directs to eardrum.
- **Middle ear** transmits energy from the eardrum to the cochlea.
- **Inner ear** stimulation of hair cells causes nerve impulses to be transmitted along the auditory nerve to the brain.

The ear's subjective judgement of sound intensity level varies with frequency of the sound. The ear needs greater stimulation at higher and lower frequencies.

Noise

Noise is defined as 'unwanted sound'. Depending upon the intensity (the energy received by the ear) and the duration (time) of exposure, an employee may suffer from acute or chronic ill-health as a result of exposure to noise. A measure of the physical harm can be estimated based on the intensity and duration of the noise.

Decibels

Noise is measured in decibels (dB). An 'A-weighting', sometimes written as 'dB(A)', is used to measure average noise levels, and a 'C-weighting', or 'dB(C)', to measure peak, impact or explosive noises. Every 3dB increase doubles the noise, so what may seem like small differences in the numbers can be quite significant.

The following gives examples of noise intensity levels:

	dB(A)
■ Threshold of human hearing	0
■ Ordinary conversation at 5m	50
■ Noisy factory	90
■ Airplane propeller at 5m	130
■ Threshold of pain	140

Types of noise ill-health effects:

- **Temporary threshold shift** will reduce the ability to hear speech at particular frequencies/pitch (known as the 4kHz dip); recovery will be rapid at first and then proceed slowly.
- **Acute tinnitus**/ringing in the ears: caused by intensive stimulation of the auditory nerves and is often experienced after attending a loud concert, so is temporary.
- **Permanent threshold shift** will reduce the ability to hear speech at particular frequencies/pitch (known as the 4kHz dip); this is irreversible.
- **Chronic tinnitus**/ringing in the ears: caused by intensive stimulation of the auditory nerves and is often experienced by professional musicians or others in constantly noisy work environments; this is permanent.
- **Acoustic trauma** caused by excessive loud noises (e.g. explosion) results in permanent damage to the ear.
- **Noise-induced hearing loss** (NIHL) is the inability to hear certain sounds and is irreversible.
- **Presbyacusis** is natural hearing loss with age.

Risk assessment/control measures for noise

Workers need to be protected from noise levels that may damage their hearing. This may be done by:

1. Identifying noise hazards (i.e. equipment and tasks that are noisy).
2. Identify those at risk (e.g. employees and any other person in the same area).

3 Evaluate levels of noise: by undertaking noise assessments.
4 Reduce noise exposure that produces risks using the hierarchy of control.
5 Record all significant findings.
6 Review risk assessment/findings as required.
7 Carry out health surveillance for those at risk of hearing damage.

The hierarchy of control for noise is outlined below.

Elimination/substitution
Is it possible to carry out the task without generating excessive noise?

For example:

- Newer models of work equipment are generally quieter and vibrate less than previous models.
- Substituting a machine using a diesel engine with one that is electrically driven.

Engineering controls
Can the development of engineering control reduce the risk?

For example:

- Controlling vibration through the use of resilient machinery mounts and flexible pipes.
- Damping – using metal or plastic to absorb energy.
- Noise absorption through the use of acoustic-absorbing ceiling baffles and screens.
- Providing sound-proof enclosures such as a hood for a printer (i.e. insulation).
- Using silencers to reduce sound energy emitted from exhaust pipes.
- Isolation through the use of rubber mounts to isolate noise.

Procedural controls
Is it possible to change the work process or reduce the exposure time?

For example:

- Job rotation – reducing exposure time of a person to a particular task.

- Introducing a programme of planned maintenance which would include lubrication of moving parts of machinery.
- Changing the process (e.g. using screws instead of rivets).

Behavioural controls
The user must take responsibility for themselves.

For example:

- Following information, instructions and training already undertaken.
- Using all protective measures, including wearing personal protective equipment (PPE) (i.e. ear plugs and earmuffs). This should be used as a last resort for noise control. It is important that consideration is given regarding the types of PPE to be used.

Health surveillance
Benefits of carrying out audiometry testing include:

- It may be used at pre-employment medicals for identifying those with existing noise-induced hearing loss.
- It can identify susceptible, vulnerable individuals.
- It provides an opportunity to train and educate employees about the prevention of hearing loss in the workplace.
- It provides evidence of the effectiveness of a hearing conservation programme.

Limitations of carrying out audiometry testing include:

- It cannot establish the extent to which hearing damage arises from occupational and non-occupational exposure.
- It is dependent on reliable subject response, operator competence and avoidance of compounding factors such as temporary threshold shift, tinnitus, etc.
- It is reactive in nature in that it only identifies hearing damage after the event.
- Resource implications – cost, time, inconvenience.
- The possibility of promoting claims by employees.

> **ACTIVITY**
>
> Review your workplace noise policy. How could it be improved?
>
> Complete a risk assessment or review your workplace noise risk assessment.

VIBRATION

Key terms to consider with vibration are: amplitude, frequency, duration and direction.

- **Amplitude** is the extent of vibration from the point of rest. Amplitude only considers the vibration in one direction, the magnitude of the vibration. Considering the vibration in both directions is a more useful indicator.
- **Frequency** of vibration is expressed in cycles per second or hertz (Hz), as used for noise measurement.
- **Duration** of vibration of exposure is the time spent by the user of the equipment.
- **Direction** of movement is along three separate axes which the tool and consequently the hand of the operator can travel, referred to as x, y and z axes. These represent up-and-down, side-to-side and backward-and-forward movements of the hand.

Types of vibration: ill-health effects

Vibration affects the body and, in extreme cases, can produce vascular disorders. It can also be responsible for nerve damage. Its effect may be localized, such as resulting from holding hand tools, or full body, such as resulting from sitting in a poorly designed driving position on vibrating machinery.

Hand/arm vibration
Examples of activities that might put a worker at risk of hand/arm vibration ill-health effects include:

- Compacting sand, concrete and aggregate.

- Cutting stone, metal and wood.
- Drilling and breaking rock, concrete and road surfaces.
- Grinding, sanding and polishing wood and stone.
- Riveting, caulking and hammering.

Exposure to this type of vibration can affect the health of the employee by causing vibration white finger, nerve damage, muscle weakness, joint damage and carpel tunnel syndrome.

Whole-body vibration

Examples of activities that might put a worker at risk of whole-body vibration ill-health effects include:

- Driving dumper trucks, off road vehicles, etc.
- Incorrect adjustment of the driver's seat.
- Poor posture while driving.

Exposure to this type of vibration can affect the health of the employee by causing back disorders (e.g. back pain and disc problems, abdominal pain, digestive disorders, urinary problems, balance issues as well as headaches and visual problems).

Risk assessment/control measures for vibration

Workers need to be protected from high levels of vibration. This may be achieved by:

1. Identifying vibration hazards (i.e. the equipment that creates vibration).
2. Identifying those at risk.
3. Evaluating levels of vibration.
4. Reducing vibration exposure that produces risks using the hierarchy of control.
5. Recording all significant findings.
6. Reviewing risk assessment/findings as required.
7. Carrying out health surveillance for those using vibration equipment.

Elimination/substitution

Is it possible to do the task without causing vibration? Newer models of work equipment are generally quieter and vibrate less than previous models.

For example:

- Substituting the tools in use with equipment with lower vibration characteristics.

Engineering controls

Can the development of engineering control reduce the risk?

For example:

- Anti-vibration handles fitted to the equipment by the manufacturer.
- Design of newer equipment creates less vibration.

Procedural controls

Is it possible to change the work process or reduce the exposure time?

For example:

- Job rotation – reducing the exposure time to a particular task.
- Reducing the exposure time of the workers by, for example, using equipment that will complete the work more quickly.
- Introducing a planned maintenance programme for tools to avoid increased vibration caused by faults or general wear.

Behavioural controls

Users must take responsibility for themselves.

For example:

- Following information, instructions and training undertaken.
- Wearing personal protective equipment such as gloves, hats and waterproofs to keep operators warm and dry – cold increases levels of risk from vibration.
- Maintaining a good blood circulation by giving up or cutting down on smoking.
- Massaging and exercising fingers during work breaks.

Health surveillance

Health surveillance should be carried out when employees are likely to be regularly exposed to excessively high levels of vibration, or where risk assessment identifies that the frequency and severity may pose a risk to health, or for employees who already have a diagnosis of hand/arm vibration.

> **ACTIVITY**
>
> Identify all the work equipment in your workplace that could cause hand/arm vibration.
>
> Identify all the work equipment in your workplace that could cause whole-body vibration.
>
> Review your workplace vibration policy. How could it be improved?
>
> Complete a risk assessment or review your workplace vibration risk assessment.

FIRST AID

The purpose of first aid is to:

- **Preserve** life.
- **Prevent** condition worsening/minimize consequences until medical help arrives.
- **Promote** recovery.
- **Provide** treatment when medical attention is not needed.

Employers should carry out an assessment of first aid needs. This will involve consideration of workplace hazards and risks, the size of the organization and other relevant factors to determine what first aid equipment, facilities and personnel should be provided. These include:

- Distribution of the workplace (is it over a large area?).
- Hazards at the workplace and type of injury that may result.
- Number of first aiders/kits currently available – are more needed?
- Past accidents within the workplace.

- Shift/holiday cover for first aiders.
- Size of organization (i.e. number of employees working and non-employees at place of work).
- Specialist training (i.e. oxygen administration or use of an automated external defibrillator AED).

In the UK there are different roles within first aid. Since 2010 they have been outlined as follows:

- First aider at work (FAW) – holding a current three-day approved qualification.
- Emergency first aider at work (EFAW) – holding a current one-day approved qualification – was previously called the appointed person.
- Appointed person – role includes looking after first-aid equipment and facilities and calling the emergency services when required. Appointed persons do not need first aid training (A FAW or EFAW can be the appointed person).

Training courses in the UK are:

- FAW: involve at least 18 hours of training and are delivered over a minimum of three days.
- FAW re-qualification: involve at least 12 hours of training and are delivered over a minimum of two days.
- EFAW: involve at least six hours of training and are delivered over a minimum of one day.

Many countries around the world have adopted this standard, including African countries such as Nigeria.

Guidance by the UK's Health and Safety Executive for the number of first aiders is as follows:

Workplace	*Number of employees*	*First aid personnel*
Low risk/hazard	Fewer than 25	At least one appointed person.
	25–50	At least one first aider trained in EFAW.
	More than 50	At least one first aider trained in FAW for every 100 employed (or part thereof).
High risk/hazard	Fewer than 25	At least one appointed person.
	25–50	At least one first aider trained in EFAW or FAW depending on the types of injuries that might occur.
	More than 50	At least one first aider trained in FAW for every 50 employed (or part thereof).

Low risk/hazard workplaces are offices, shops, libraries, etc.

High risk/hazard workplaces are light engineering and assembly work, food processing, warehousing, extensive work with dangerous machinery or sharp instruments, construction, chemical manufacturing, etc.

In 2011 a new contents list for first aid kits was introduced (British Standards-8599).

Contents	*Small*	*Medium*	*Large*	*Travel*
First aid guidance leaflet	1	1	1	1
Contents list	1	1	1	1
Medium dressing (12cm x 12cm) (sterile)	4	6	8	1
Large dressing (18cm x 18cm) (sterile)	1	2	2	1
Triangular bandage (single use) (90cm x 127cm)	2	3	4	1
Safety-pins (assorted) (minimum length 2.5cm)	6	12	24	2
Eye pad dressing with bandage (sterile)	2	3	4	0
Wash-proof assorted plasters	40	60	100	10
Moist cleansing wipes	20	30	40	4
Micro-porous tape (2.5cm x 5m or 3m for Travel Kit)	1	1	1	1
Nitrile gloves (pairs)	6	9	12	1
Finger dressing with adhesive fixing (3.5cm)	2	3	4	0
Mouth-to-mouth resuscitation device with valve	1	1	2	1
Foil blanket (130cm x 210cm)	1	2	3	1
Eyewash (250ml)	0	0	0	1
Burn relief dressing (10cm x 10cm)	1	2	2	1
Universal shears (suitable for cutting clothing)	1	1	1	1
Conforming bandage (7.5cm x 4m)	1	2	2	1

- Small first aid kit suitable for fewer than 25 people in a low risk environment and fewer than five people in a high risk environment.
- Medium first aid kit suitable for 25 to 100 people in a low risk environment and five to 25 people in a high risk environment.
- Large first aid kit suitable for more than 100 people in a low risk environment and more than 25 people in a high risk environment.

ACTIVITY

Identify who are the first aiders, and the level at which they are qualified, in your workplace. Taking the above information into account, consider if the current policy is suitable.

KEY NOTE

As you have worked through these key health and safety topics you should be able to remember parts as you read your course book, or vice-versa.

This book has repeated parts of fire and manual handling, have you noticed? This is to show you how the brain begins to remember information.

RECALL

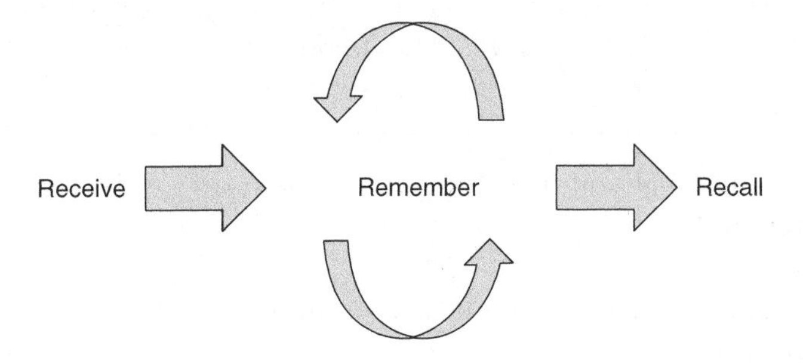

Once you have completed your revision and are able to remember the information you have received, you will need to get ready to recall it in order to sit your exam successfully.

The most important part of recalling the information for an exam is that you focus on what the exam requires.

This section of the book will endeavour to look specifically at answering both the multiple choice questions (MCQ) and the written exam for professional certificate qualifications and CIEH Level 4 awards.

PREPARATION FOR SITTING AN MCQ OR WRITTEN EXAM

The awarding body or your course provider may send you an exam slip that you will need to bring with you to a professional certificate qualification exam. You may also be required to bring photographic identification (e.g. a passport or driving licence). The invigilator of your exam may be required to check your identification – if you are unable to provide any you could be prevented from taking the exam!

You will be provided with details about your exam by your course provider, including the date, time, venue, etc. You should ensure that you arrive in plenty of time for your exam.

CASE STUDY

Paul preferred to arrive at least one hour before his exam. He would find somewhere to sit down, have a cup of coffee and quickly read over his notes.

It is imperative that you have everything you need for the exam. This is likely to include:

- Pens (at least two blue or black ball-point).
- Small ruler.
- 2B or HB pencils/sharpener/eraser (for sketches or to answer MCQs).
- Clear plastic pencil case.
- Bottle of water.

Do not take highlighter pens or correction fluid into the exam. These are not to be used on exam papers.

It is unlikely that you will be allowed to take any course books, notes, etc. into the examination room unless you are undertaking an open book exam. The invigilator will instruct you as to where you can store your personal belongings during the exam. It is possible that your seat will have been previously allocated or you may be able to choose where to sit. This will become clear upon entering the examination room. The desks in the examination room should be set out in rows, all facing the same direction, with about 2 metres between each desk. This is to ensure that you cannot see other students' material and that they cannot see yours. Strict silence is required in exams. You must not use any electronic communication equipment (mobile phones, MP3s, pagers, etc.).

Any breach of examination procedures may result in your examination paper being made void and you becoming unable to take future exams with that particular awarding body.

MULTIPLE CHOICE QUESTIONS

ANSWERING MULTIPLE CHOICE QUESTIONS

The number of questions and the complexity increases in line with the level of the qualification. Multiple choice questions (MCQs) set by the awarding body should be based on the learning outcomes in the respective syllabi.

Health and safety passport schemes that use MCQs include:

Client/Contractor National Safety Group	http://www.ccnsg.com
Safety Passport Alliance	http://www.safetypassports.co.uk
Construction Skills	http://www.cskills.org

Health and safety courses that use MCQs include:

- CIEH and HABC both offer a Level 1 award (half-day programme).
- CIEH and HABC both offer a Level 2 award (one-day programme).
- CIEH Level 3 (three-day programme).
- NEBOSH Level 2 (18 guided learning hours). Award also requires the completion of an assignment.
- HABC Level 3 award (30 guided learning hours).
- HABC Level 4 award (40 guided learning hours).

Note: This list in not exhaustive.

EXAMPLES

KEY NOTE

The following MCQ examples have been written to indicate the styles of questions used by all awarding bodies delivering at differing health and safety qualifications levels. Any resemblance to actual questions set by any awarding body is purely coincidental.

Each of the following examples has been developed to give you the essential study skills to be able to take an MCQ assessment, after prior study of the course material.

EXAMPLE 1

The colour and shape for a HAZARD sign is normally:

A Yellow and black, triangle
B Blue and white, circle
C Red and white, circle
D Green and white, rectangle

This question requires you to be able to identify the different types of signs and give the correct answer; a hazard sign is **A**: Yellow and black, triangle. Other questions that may have been used to test your knowledge could have been:

- The colour and shape for a MANDATORY sign is normally:
- The colour and shape for a PROHIBITION sign is normally:
- The colour and shape for a SAFE CONDITION sign is normally:

Alternatively, the question could have been phrased as:

EXAMPLE 2

What does a yellow and black triangle sign mean?

A Safe condition
B Mandatory
C Prohibition
D Hazard

The answer is **D:** Hazard.

To be able to answer any question about signs, it is important that you familiarize yourself with them.

EXAMPLE 3

What action would you take in the event of discovering a fire?

A Tell your supervisor
B Tackle the fire by yourself
C Raise the alarm and go to the assembly point
D Go home

This question tests the knowledge of what action should be taken in the event that you discover a fire. You could give any one of the four answers; however, the correct action in this instance would be **C**: Raise the alarm and go to the assembly point.

EXAMPLE 4

Which of the following is **not** a symptom and sign of hearing loss?

A Tinnitus
B Difficulty hearing conversations
C Chest pains
D Your family complains about the volume on the television being too loud

This question tests your knowledge of noise. You may not know what tinnitus is (it should be covered on your course); however, **C** has nothing do with noise, so this must be the correct answer! Tinnitus is ringing, whistling, buzzing or humming in the ears, which can be either permanent or temporary.

EXAMPLE 5

Which of the following is correct?

A A hazard is a result of an accident
B A hazard is the likelihood and severity of an accident occurring
C A hazard introduces controls to prevent a risk
D A hazard is anything that may cause harm

This question requires you to read carefully the options available, and select the correct statement: **D**, A hazard is anything that may cause harm (e.g. chemicals, electricity, working from ladders or manual handling in an unsafe manner). A process of elimination and logically using prior knowledge may be employed with this type of question.

First, hazards can cause accidents, they don't result from them, and therefore **A** is incorrect. Second, remember the definitions for risk: **B** is a definition of risk and is therefore incorrect. Third, hazards don't introduce controls but can result in controls having to be introduced, and therefore **C** is incorrect. As a result **D** can be the only correct answer.

EXAMPLE 6

Which is the first stage of the five steps to risk assessment?

A Record significant findings and implement them
B Identify the hazards
C Evaluate the hazards
D Review and update as necessary

This question requires you to recall the five steps of risk assessment, which are as follows:

1 Identify the hazards.
2 Identify the people who might be harmed and how.
3 Evaluate the risk and decide on precautions.
4 Record and implement significant findings.
5 Review and update as necessary.

The answer is **B**: Identify the hazards; however, note that answer **C**: Evaluate the hazards is very similar. Make sure you read the question carefully. To help with these types of questions, first put them in a logical order: you cannot evaluate, record or review until you have first identified the hazard, and therefore **B** is the only correct answer.

STRATEGIES FOR ANSWERING MULTIPLE CHOICE QUESTIONS

For MCQs that require you to answer a direct question (Examples 1, 2, 3 and 6 above), your approach should be to read the question and then, without looking at the four possible choices, to think what it could be recalling from your prior knowledge. If, when you look at the four choices, your 'answer' is one of the four, you are most likely to be correct.

For MCQs that don't pose a direct question (Example 4 and 5 above), the approach to answering them is slightly different. Read the question and then read each possible answer. Once you have done this, read each answer carefully followed by the question.

EXAMPLE 4 (from above, reworked)

A Tinnitus
~~Which~~ is a symptom and sign of hearing loss. ✗
B Difficulty hearing conversations
~~Which~~ is a symptom and sign of hearing loss. ✗
C Chest pains
~~Which~~ is *not* a symptom and sign of hearing loss. ✓
D Your family complains about the volume on the television being too loud
~~Which~~ is a symptom and sign of hearing loss ✗

The only true statement is **C**, therefore **C** is the correct answer.

If you are unsure about two or more, then you will need to ask yourself: which is the most appropriate? Some answers may appear to sound correct, but once you have looked at the other possible answers it should become clear which is *most* likely to be the correct answer. If you are completely unsure, leave the question and move on to the next. Once you have answered all of the other questions go back and reattempt the MCQ you were initially unsure about.

KEY NOTE

If you have to write your answers on a separate piece of paper make sure that your marks align with the correct question number. It is easy to miss out a question and to start to place marks against the wrong questions.

Once you have finished all the questions, read through *all* your answers and double-check that you have not made a mistake and have answered all the questions.

CASE STUDY

How not to answer MCQs/some less than sound approaches to answering MCQs:

If she did not know the answer confidently Kara thought that the correct possible answer would always be C.

If she did not know the answer confidently Emma would select the letter she had used the least up until that point.

MOCK MULTIPLE CHOICE QUESTIONS

KEY NOTE

The following questions have been developed to show the types of questions that could appear on award qualifications, ranging from Level 1 to Level 4, from any of the awarding bodies, and similarities are purely coincidental.

1 Health is:

A The protection of the bodies and minds of people from illness resulting from the materials, processes or procedures used in the workplace.
B The protection of people from physical injury resulting from the materials, processes or procedures used in the workplace.
C The potential of a substance, activity or process to cause harm.
D The likelihood and severity of a substance, activity or process to cause harm.

2 Safety is:

A The protection of the bodies and minds of people from illness resulting from the materials, processes or procedures used in the workplace.
B The protection of people from physical injury resulting from the materials, processes or procedures used in the workplace.
C The potential of a substance, activity or process to cause harm.
D The likelihood and severity of a substance, activity or process to cause harm.

3 Moral reasons for health and safety include:

A Reduces workplace accidents and reduces work-related ill-health and occupational diseases.
B Employer having a duty of care in criminal and civil law.
C The direct and indirect costs of an accident.
D Reduces workplace accidents, the employer having a duty of care and the costs of an accident.

4 Legal reasons for health and safety include:

A Reduces workplace accidents and reduces work-related ill-health and occupational diseases.
B Employer having a duty of care in criminal and civil law.
C The direct and indirect costs of an accident.
D Reduces workplace accidents, the employer having a duty of care and the costs of an accident.

5 Financial reasons for health and safety include:

A Reduces workplace accidents and reduces work-related ill-health and occupational diseases.
B Employer having a duty of care in criminal and civil law.
C The direct and indirect costs of an accident.
D Reduces workplace accidents, the employer having a duty of care and the costs of an accident.

6 Costs of poor health and safety at work include:

A Increased profits, high staff turnover, legal costs.
B Reduced profits, high staff turnover, legal costs.
C Good reputation, low staff turnover, increased profits.
D Poor reputation, low staff turnover, reduced profits.

7 Benefits of good health and safety at work include:

A Increased profits, high staff turnover, legal costs.
B Reduced profits, high staff turnover, legal costs.
C Good reputation, low staff turnover, increased profits.
D Poor reputation, low staff turnover, reduced profits.

8 Benefits of good health and safety at work are developed from good:

A Communication, cooperation, competence and control.
B Reduced profits, high staff turnover, legal costs.
C Good reputation, low staff turnover, increased profits.
D Poor reputation, low staff turnover, reduced profits.

9 Main factors that affect health and safety include:

A Environmental, human and occupational.
B Management, employees and self-employed.
C Job, task and individual.
D Statement of intent, organization and arrangements.

10 Primary causes of accidents in the workplace are:

A Human error.
B The public.
C Acts of God.
D Good management.

11 An accident may be described as:

A Being concerned with those illnesses or physical and mental disorders that are either caused or triggered by workplace activities.
B An unplanned event that results in injury or ill-health for people, or damage, or loss to property, plant, materials or the environment, or loss of a business opportunity.
C Any incident that could have resulted in an accident in the workplace.
D A 'near miss' which could have led to serious injury or loss of life.

12 A near miss is:

A Concerned with those illnesses that are either caused or triggered by workplace activities.
B An unplanned event that results in injury, damage or loss to property, plant or materials.
C Any incident that could have resulted in an accident.
D A 'near miss' which could have led to serious injury or loss of life.

13 A dangerous occurrence is:

A Concerned with those illnesses that are either caused or triggered by workplace activities.
B An unplanned event that results in injury, damage or loss to property, plant or materials.

C Any incident that could have resulted in an accident.
D A 'near miss' which could have led to serious injury or loss of life.

14 Ill-health is:

A Concerned with those illnesses that are either caused or triggered by workplace activities.
B An unplanned event that results in injury, damage or loss to property, plant or materials.
C Any incident that could have resulted in an accident.
D A 'near miss' which could have led to serious injury or loss of life.

15 Human factors that contribute to accidents and ill-health in the workplace are:

A Associated with the job/task (e.g. manual handling or the use of equipment).
B Skill-based errors, mistakes and violations.
C Associated with the area of work (e.g. safe access and egress, noise, lighting, etc.).
D Visitors, trespassers and the general public on site.

16 Occupational factors that contribute to accidents and ill-health in the workplace are:

A Associated with the job/task (e.g. manual handling or the use of equipment).
B Skill-based errors, mistakes and violations.
C Associated with the area of work (e.g. safe access and egress, noise, lighting, etc.).
D Visitors, trespassers and the general public on site.

17 Environmental factors that contribute to accidents and ill-health in the workplace are:

A Associated with the job/task (e.g. manual handling or the use of equipment).
B Skill-based errors, mistakes and violations.

C Associated with the area of work (e.g. safe access and egress, noise, lighting, etc.).
D Visitors, trespassers and the general public on site.

18 Key elements of a health and safety policy are:

A Statement of intent, organization and arrangements.
B Moral, legal and financial.
C Policy, organizing, planning and implementing, reviewing performance and auditing.
D Human, occupational, environmental.

19 A hazard is:

A The protection of the bodies and minds of people from illness resulting from the materials, processes or procedures used in the workplace.
B The protection of people from physical injury resulting from the materials, processes or procedures used in the workplace.
C The potential of a substance, activity or process to cause harm.
D The likelihood and severity of a substance, activity or process to cause harm.

20 A risk is:

A The protection of the bodies and minds of people from illness resulting from the materials, processes or procedures used in the workplace.
B The protection of people from physical injury resulting from the materials, processes or procedures used in the workplace.
C The potential of a substance, activity or process to cause harm.
D The likelihood and severity of a substance, activity or process to cause harm.

21 The first stage of a risk assessment is to:

A Identify hazards.
B Identify those who may be harmed.

C Evaluate risk by considering existing precautions.
D Record the findings.

22 To be competent in carrying out a risk assessment, you must:

A Be a manager or supervisor.
B Hold a Level 3 health and safety qualification.
C Have worked in that area for over three years.
D Have the necessary experience and qualifications.

23 A systematic approach should be used to eliminate and then substitute a hazard before considering other measures such as reducing exposure, and lastly, using personal protective equipment. This is called a:

A Management control.
B Hierarchy of control.
C Environmental control.
D Health and safety control.

24 Using personal protective equipment as a risk control measure should be seen as:

A The first choice.
B The last resort.
C Least cost-effective.
D Most cost-effective.

25 Benefits of consulting employees about health and safety include:

A Increases the number of meetings, gains experience from a range of people in the organization and develops trust between employer and employees.
B To assess risks, charge employees for attending meetings and develop trust between employer and employees.
C To assess risks, gain experience from a range of people in the organization and pass accountability from employer to employee.
D To assess risks, gain experience from a range of people in the organization and develop trust.

26 Signs that indicate a HAZARD should normally be:

A Yellow and black.
B Blue and white.
C Red and white.
D Green and white.

27 Signs that indicate a MANDATORY instruction should normally be:

A Yellow and black.
B Blue and white.
C Red and white.
D Green and white.

28 Signs that indicate a PROHIBITION instruction should normally be:

A Yellow and black.
B Blue and white.
C Red and white.
D Green and white.

29 Signs that indicate a SAFE condition should normally be:

A Yellow and black.
B Blue and white.
C Red and white.
D Green and white.

30 Factors to be considered when carrying out a manual handling risk assessment are:

A That the load must not be over 25kg.
B Task, load, environment, individual capability and other factors.
C Job, task and individual.
D Environmental, human and occupational.

31 The requirements for first aid in the workplace are to:

A Preserve life, provide training on CPR, purchase first aid kits.
B Promote recovery, prevent the condition worsening, purchase first aid kits.

C Preserve life, promote recovery, prevent the condition from worsening.
D Prevent the condition from worsening, provide training on CPR, promote recovery.

32 Four routes of harmful agents entering the body are:

A Inhalation, electrocution, injection, ingestion.
B Inhalation, absorption, radiation, ingestion.
C Inhalation, absorption, injection, hearing.
D Inhalation, absorption, injection, ingestion.

33 Acute health effects:

A Occur soon after the exposure and are often of short duration.
B Develop over time.
C Caused by inhalation, absorption, injection or ingestion.
D Caused by environmental, human and occupational factors.

34 Chronic health effects:

A Occur soon after the exposure and are often of short duration.
B Develop over time.
C Caused by inhalation, absorption, injection or ingestion.
D Caused by environmental, human and occupational factors.

35 Accident investigation is an example of:

A Reactive measuring.
B Proactive measuring.
C Causes of workplace accidents.
D Elements of the health and safety policy.

36 Audits, surveys, inspections, tours and samples are all examples of:

A Reactive measuring.
B Proactive measuring.
C Causes of workplace accidents.
D Elements of a health and safety policy.

37 Key issues in a workplace inspection include:

A Accident investigation.
B Premises, plant and substances, procedures and people.
C Management review.
D Staff appraisals.

38 An example of the activities to be undertaken during a safety tour would be:

A To check on issues such as wearing PPE and housekeeping.
B By focusing on a particular activity (e.g. manual handling).
C A comprehensive and independently executed examination of all aspects of an organization's health and safety performance against set objectives.
D By comparing the organization's safety performance against other, similar organizations, or from one department to another.

39 An example of the activities to be undertaken during a safety survey would be:

A To check on issues such as wearing PPE and housekeeping.
B By focusing on a particular activity (e.g. manual handling).
C A comprehensive and independently executed examination of all aspects of an organization's health and safety performance against set objectives.
D By comparing the organization's safety performance against other, similar organizations, or from one department to another.

40 An example of the activities to be undertaken during a safety audit would be:

A To check on issues such as wearing PPE and housekeeping.
B By focusing on a particular activity (e.g. manual handling).
C A comprehensive and independently executed examination of all aspects of an organization's health and safety performance against set objectives.
D By comparing the organization's safety performance against other, similar organizations, or from one department to another.

MOCK MCQ ANSWERS

1 A
2 B
3 A
4 B
5 C
6 B
7 C
8 C
9 A
10 A
11 B
12 C
13 D
14 C
15 B
16 A
17 C
18 A
19 C
20 D
21 A
22 D
23 B
24 B
25 D
26 A
27 B
28 C
29 D
30 B
31 C
32 D
33 A
34 B
35 A
36 B
37 A
38 A
39 B
40 C

WRITTEN EXAMS

ANSWERING EXAM QUESTIONS

Written exam questions provide an opportunity to assess your knowledge and understanding.

COMMON MISTAKES MADE BY STUDENTS DURING EXAMINATIONS

NEBOSH EXAMINERS' REPORT

Each NEBOSH examiners' report includes details of common pitfalls that are beneficial for all health and safety students to be aware of:

- Many candidates fail to apply the basic principles of examination technique, and for some candidates this means the difference between a pass and a referral.
- In some instances, candidates are failing because they do not attempt all the required questions or are failing to provide complete answers. Candidates are advised to always attempt an answer to a compulsory question, even when the mind goes blank. Applying basic health and safety management principles can generate credit-worthy points.
- Some candidates fail to answer the question set and instead provide information that may be relevant to the topic but is irrelevant to the question, and cannot therefore be awarded marks.
- Many candidates fail to apply the command words (also known as action verbs; for example, describe, outline, etc.). Command words are the instructions that guide the candidate on the depth of answer required. If, for instance, a question asks the candidate to 'describe' something, then few marks will be awarded to an answer that is an outline.
- Some candidates fail to separate their answers into the different subsections of the questions. These candidates could gain marks for the different sections if they clearly indicated which part of the question they were answering (e.g. by using the numbering from the question in their answer). Structuring their answers to

address the different parts of the question can also help in logically drawing out the points to be made in response.

- Candidates need to plan their time effectively. Some candidates fail to make good use of their time and give excessive detail in some answers, leaving insufficient time to address all the questions.
- Candidates should also be aware that examiners cannot award marks if handwriting is illegible.

Source: From March 2010 NEBOSH Examiners' Report (NGC1), accessed online, July 2012

CIEH MODERATORS' REPORT

This report also provides general advice that can be used by all health and safety students:

- Many candidates fail to understand the requirements of some questions as a result of misinterpreting the question.
- Candidates must be fully conversant with the whole syllabus and the scope of the subjects within it, rather than only focusing on specific topics.
- Candidates who only make lists are not 'describing', 'explaining' or 'outlining' as specifically requested in the question and fail to provide the relevant detail. This severely limits the number of marks that can be awarded.
- Candidates need to be able to answer questions based on any part of the syllabus (as it may be applied generally in a workplace), so it is essential that they address and revise every element in their course programme. Candidates' answers sometimes appear limited and fail to address the full scope of questions or identify wider issues other than subject basics.
- Candidates must be reminded to read examination questions carefully in order to understand what is being asked of them. It is still apparent that some candidates do not read questions properly and as a result waste both time and marks by giving inappropriate and irrelevant answers.
- There are too many occasions when candidates provide one-word or very short answers, which do not provide sufficient depth or detail at this level.

- Candidates are required to give reasoned answers to questions and should ensure that they have sufficient knowledge to respond in depth. Bullet-pointed lists do not generally provide sufficient depth and breadth of information and tend to be restrictive. Unless the question specifically requests it, a bullet-pointed list will not be eligible for the full range of marks; neither will vague or generalized statements such as 'following good practice'.
- Writing copiously on a subject without considering the finer points of the question is not an effective practice. It leads to digression and unnecessary/irrelevant information, which will not gain any marks.
- Some candidates repeat similar points, or the same point by means of an alternative phrase, when additional marks are available for citing different examples. Duplicated material can only be credited once.
- On occasion, candidates' handwriting is difficult to read, and sometimes it cannot be deciphered at all – if a candidate's answers are not clear, marks will be limited.

Source: From CIEH Annual Moderators' Report 2010–2011, accessed online, July 2012

UNDERSTANDING THE ACTION VERBS/COMMAND WORDS

Each exam question will use an action verb/command word as part of the question. These are designed to test the students' knowledge or understanding of a particular area of study.

Questions set by the awarding body should be based on the learning outcomes in the respective syllabus.

Action verbs/command words are used in health and safety qualifications at award, certificate and diploma level. They are also used for GCSE/A-level and other courses.

CASE STUDY

From experience in the classroom:

Referring to the action verbs, some students have stated that the questions on health and safety papers are awkward to decipher and seem like jargon. This thinking has become a mindset among health and safety students and, unfortunately, some trainers. Action verbs are not used solely in health and safety examinations; they are also used in many other subject areas.

I try to explain how the questions are structured – that they are based on the syllabi for each course and that they are simply testing the students' knowledge and understanding of what is on the syllabus. I explain that the questions use action verbs to help students maximize their marks.

USING ACTION VERBS/COMMAND WORDS IN QUESTIONS

Question

Identify different action verbs that are used in health and safety exam questions and **outline** the meaning of each (20); OR

List different action verbs that are used in health and safety exam questions and **state** the meaning of each (20).

Answer

Action verbs used in health and safety qualifications include:

Action verb	*Meaning*
1 Give	Provide specific information without explanation.
2 Identify	Select and name something.
3 List	Provide a list without any further explanation.
4 Name	Provide the name without any further explanation.
5 Define	Provide a recognized/common definition.
6 Describe	Provide an in-depth explanation.
7 Explain	Provide a clear account of, or reasons for.
8 Outline	Provide the most important features.
9 State	Provide a brief explanation/express a view.
10 Compare and contrast	Provide an explanation of similarities and differences.

Note: This list in not exhaustive.

The following examples of action verbs are based on manual handling activities in the workplace. They have not been directly taken from questions set by the awarding bodies. Some questions may fall outside the learning outcomes of the awarding bodies and therefore may not be suitable as exam questions, but they have been used here so that you will be able to understand more thoroughly the action verbs usage.

Question	*Possible answer/part answer*
1. **Give** examples of those who may be involved in assessing an employee who has injured their back at work	■ Occupational physiotherapists. ■ Occupational therapists. ■ Occupational nurse. ■ Their General Practitioner.
2. **Identify** types of manual handling injuries in the workplace	■ Prolapsed disc. ■ Sprained ligament. ■ Strained muscle. ■ Dislocated joint. ■ Fractured bone. ■ Cuts. ■ Hernia. ■ Crushing injuries.
3. **List** four areas that are assessed during a manual handling risk assessment	■ Task. ■ Load. ■ Individual. ■ Environment.
4. **Name** the five regions of the spine	■ Cervical. ■ Thoracic. ■ Lumbar. ■ Sacral. ■ Coccygeal.
5. **Define** the term 'manual handling'	'Any transport or support of a load (including the lifting, putting down, pushing, pulling, carrying or moving thereof) by hand or by bodily force'
6. **Describe** the duties placed on an employer under the Manual Handling Operations Regulations 1992	■ Avoid the need for hazardous manual handling, so far as is reasonably practicable. ■ Assess the risk of injury from any hazardous manual handling that cannot be avoided. ■ Reduce the risk of injury from hazardous manual handling, so far as is reasonably practicable.

(*Continued*)

Question	Possible answer/part answer
7. **Explain** the term Upper Limb Disorders (ULDs)	■ Include aches, pains, tension and disorders involving any part of the arm from fingers to shoulder, or the neck. ■ Include problems with the soft tissues, muscles, tendons and ligaments, along with the circulatory and nerve supply to the limb. ■ Are often caused or made worse by work.
8. **Outline** the correct technique for manual handling	■ Think before lifting/handling. ■ Keep the load close to the waist. ■ Adopt a stable position and gain a good hold. ■ At the start of the lift, slight bending of the back, hips and knees is preferable to fully flexing the back (stooping) or fully flexing the hips and knees (squatting). ■ Don't flex the back any further while lifting. ■ Keep the head up when handling. ■ Move smoothly. ■ Don't lift or handle more than can be easily managed. ■ Put down, then adjust.
9. **State** the reasons why a manual handling incident should be investigated	■ Prevent manual handling incident recurring. ■ Identify weakness in manual handling risk assessment/policy. ■ To determine cause of the incident.
10. **Compare and contrast** sprains and strains	■ Both can result from manual handing injuries **A sprain is an injury to ligaments.** **A strain is an injury to muscle or tendon tissue.** ■ For example: **Sprain to wrist.** **Strain to lower back.**

Please note that the answers given are not the only possible answers; for example, there could be others than those listed which are involved in assessing an employee who has injured their back at work.

As you will see from the above:

- **Give** examples of; **identify**; **list** and **name** required short answers, using only a few words.
- **Define**; **describe**; **explain**; **outline**; **state** and **compare and contrast** required sentences to enable you to give a full answer.

Some questions may require you to provide an example to clarify your answer. This applies to all action verbs.

If you use the words *information, instruction, training* and *supervision* as part of an answer, make sure the examiner knows what you mean. For example:

- Provide information of the chemicals used by employees.
- Ensure the operator of the computer has been instructed on how to set up the chair correctly.
- Training should be given to the chainsaw operator (e.g. attend a one-day course prior to operating).
- Ensure that employees are supervised – especially newly trained or young persons.

If your answer includes PPE (personal protective equipment), make sure you give an example:

- Gloves should be worn to prevent employees from cutting their hands.
- Life vests should be worn if working on the water (in a boat) or over water (in a cherry picker).

KEY NOTE

If you only write down IITS or information, instruction, training and supervision without saying what type, this is unlikely to gain the full marks available.

The following questions, about hard hats, have been developed for you to gain an understanding of the amount of information that is needed to gain maximum marks in any question.

See if you can answer the following:

1. **Give** an example of a workplace where a hard hat should be worn.
2. **Identify** the sizes of hard hats available.
3. **List** the colours of hard hat available.
4. **Name** the makes/manufacturers of hard hats.
5. **Define** a hard hat.
6. **Describe** the safety features of a hard hat.
7. **Explain** the purpose of wearing a hard hat.
8. **Outline** the benefits of wearing a hard hat.

9 **State** the disciplinary action to be taken if an employee is not wearing his or her hard hat in a designated area.
10 **Compare** and **contrast** the use of a bump-cap and hard hat.

Your answers should reflect the following:

- **Give** examples of; **identify**; **list** and **name** require short answers, using only a few words.
- **Define**; **describe**; **explain**; **outline**; **state,** and **compare and contrast** require sentences to enable you to give a full answer.

Note that **identify** sometimes requires more than one or two words; see Example 2, Five steps (below) for details.

> **KEY NOTE**
>
> If you only write PPE without giving an example, this is unlikely to gain the full marks available.

KNOWLEDGE AND UNDERSTANDING

The questions set will test both your knowledge and your understanding of a topic within health and safety. Generally speaking, the examination action verbs fall into two categories, but there will always be exceptions to this:

Knowledge	■ Give examples of.
	■ Identify.
	■ List.
	■ Name.
	■ Define.
Understanding	■ Describe.
	■ Explain.
	■ Outline.
	■ State.
	■ Compare and contrast.

The number of marks available will indicate how much information is needed. Generally, each mark requires a separate piece of information; for example, a 4-mark question will require four pieces of information.

EXEMPLAR WRITTEN EXAM QUESTIONS

> **KEY NOTE**
>
> The following health and safety questions have been written to indicate the styles of questions used by all awarding bodies at all levels. Any resemblance to actual questions set by any of the awarding bodies is purely coincidental. All answers have been devised to provide examples of good and bad practice/exam technique.

The phrase commonly used by joiners is: measure twice, cut once. This principle may also be applied to exam questions. Before you start to write down your answer, read the question fully, and then reread the question a second time before committing pen to paper.

EXAMPLE 1: MISREADING OR MISUNDERSTANDING THE QUESTION

Identify the hazards when using an electrical piece of equipment (2 marks).

The keyword in this question is *hazard* – it is often the case that students either misread *hazard* as risk or as control when answering this type of question.

The hazard is something with the potential to cause harm, so the answer would include:

- Damaged cables.
- Overloaded sockets.

If the question was: **Identify** the risks when using an electrical piece of equipment (2 marks) the answer would include the following:

Risk is the likelihood that a specified, undesired event will occur due to the realization of a hazard; or risk = severity x likelihood – so the answer would include:

- Risk of electrocution/electric shock to operator.
- Risk of fire or overheating of equipment.

If the question was: **Outline** the controls when using an electrical piece of equipment (2 marks) the answer would include the following:

The control measures are put in place to eliminate or reduce the risk – so the answer would include:

- Ensure equipment is regularly maintained (e.g. portable appliance testing).
- Ensure that extension cables are not used fully wound up.

KEY NOTE

Before you answer a question underline the important words in the question.

EXAMPLE 2: FIVE STEPS

Some questions are, or should be, straightforward for health and safety students to answer:

- What are the five steps to risk assessment? (5 marks); OR
- **Identify** the five steps to risk assessment (5 marks); OR
- **State** the five steps to risk assessment (5 marks).

The Health and Safety Executive (HSE) document *Five Steps to Risk Assessment*, which may be found at http://www.hse.gov.uk/pubns/indg163.pdf, explains the principles of the five steps.

Identify the five steps to risk assessment (5 marks):

✓ 1 Identify the hazards.
✓ 2 Identify the people who might be harmed and how.
✓ 3 Evaluate the risk and decide on precautions.
✓ 4 Record significant findings and implement them.
✓ 5 Review and update as necessary.

5/5

To gain the maximum marks you should be able to write down the five steps, as shown above. Note that an **Identify** question sometimes requires more than one or two words.

This question could be expanded on a written paper to:

- **Identify** the five steps of risk assessment and **give** an example of each stage (10 marks); OR
- **State** the five steps of risk assessment and **describe** what is involved at each stage (10 marks).

This would test your understanding rather than just your knowledge of the five steps of risk assessment.

EXAMPLE 3: MANUAL HANDLING INJURIES

A question to test your knowledge of manual handling injuries associated with the workplace could be worded as follows:

- **Give** examples of manual handling injuries associated with the workplace (4 marks); OR
- **Identify** types of manual handling injuries associated with the workplace (4 marks); OR
- **List** different types of manual handling injuries associated with the workplace (4 marks); OR
- **Name** the types of manual handling injuries associated with the workplace (4 marks); OR
- **What** are the types of manual handling injuries that could occur in the workplace? (4 marks).

Possible answers to this question are:

- Slipped/prolapsed disc.
- Sprain of ligament or tendon.
- Strain of muscle.
- Dislocation.
- Fracture.
- Cuts/abrasions.
- Hernia/ruptures.
- Crushing/impact injuries.

It is likely that the question would have been worth four marks, even though eight possible answers are given above. The reason would be to give you the maximum opportunity to achieve all the marks available.

Give examples of manual handling injuries in the workplace (4 marks):

✗ 1 Bad back.
✓ 2 Slipped disc.
(2) 3 Prolapsed disk.
✓ 4 Strain to muscles in lower back.
2/4

The above question only scored two out of a possible four marks. By looking at each potential answer it will be possible to understand what the examiner is looking for and why two marks were lost.

Reasons for marks awarded/not awarded:

1 **Bad back**
It is true that as a result of a manual handling injury someone may suffer a bad back; however, this is not an injury in itself, it is the result of an injury such as prolapsed disc and therefore no mark was awarded.
2 **Slipped disc**
Correct: one mark. At this level a slipped disc could be awarded, as this is a 'layman's' term to describe a prolapsed disc.
3 **Prolapsed disk**
Correct, but no mark. The reason is not because of the incorrect spelling of 'disc'. Examiners will not penalize spelling errors; however, if the handwriting is poor the examiner may not be able to read it and will

therefore not be able to award marks. The reason why this did not get a mark is because the answer above (slipped disc) is the same answer.

4 **Strain muscles in lower back**
Correct: one mark. This is a good answer, as it explains where the strained muscle is. A manual handling injury could result in a strained muscle (for example, in the lower back) or a sprained ligament (for example, in the wrist or ankle).

The following answers are also correct for strain, but only one mark can be awarded per category of injury:

- Strained muscle.
- Pulled muscle.
- Torn muscle.
- Strain of the lower back.

EXAMPLE 4: MANUAL HANDLING ASSESSMENT (A–D)

Questions that use the action verbs **describe**; **explain**; **outline**; **state**, and **compare and contrast** require more information than the knowledge-based questions (**give** examples of; **identify**; **list**; **name** and **define**).

The four factors that should be assessed when carrying out manual handling risk assessments are:

- Task.
- Load.
- Individual.
- Environment.

To test your understanding, one of the following questions may be used:

- **Describe** the factors to be considered when conducting a manual handling risk assessment (8 marks); OR
- **Explain** the factors to be considered when conducting a manual handling risk assessment (8 marks); OR
- **Outline** the factors to be considered when conducting a manual handling risk assessment (8 marks); OR
- **State** the factors to be considered when conducting a manual handling risk assessment (8 marks).

Possible answers to this question will include reference to:

The tasks:

- Holding loads away from trunk.
- Twisting.
- Stooping.
- Reaching upward.
- Large vertical movement.
- Long carrying distances.
- Strenuous pushing or pulling.
- Unpredictable movement of loads.
- Repetitive handling.
- Insufficient rest or recovery.
- A work rate imposed by a process.

Individual capability:

- Require unusual capability.
- Hazard those with a health problem.
- Hazard those who are pregnant.
- Call for special information/training.

The loads:

- Heavy.
- Bulky/unwieldy.
- Difficult to grasp.
- Unstable/unpredictable.
- Intrinsically harmful (e.g. sharp/hot).

The environment:

- Constraints on posture.
- Poor floors.
- Variations in level.
- Hot/cold/humid conditions.
- Strong air movements.
- Poor lighting conditions.

Source: Adapted from Manual Handling Operations Regulations 1992 (as amended), Guidance on Regulations, HSE Document L23

The answer sheet (above) shows over 20 possible answers. Note that the four subheadings are unlikely to be awarded as marks on their own, but may be used by you to develop your answer.

The following are four possible answers given by students A–D and their awarded marks with explanatory remarks.

Student A

Outline factors that should be assessed during a manual handling risk assessment (8 marks):

1 About the task.
2 Individual factors.
3 Loads involved.
4 Working environment.

2/8

Student A simply identified the four areas that should be assessed but did not **outline** any additional information/details.

Since Student A provided some knowledge, it is only fair that some marks are awarded. It's not worth a mark per answer; so only two marks can be given in total.

Please note that if the question was **Identify/List** or **Name** the factors that should be assessed during a manual handling risk assessment then 4 marks would have been awarded. Since only four possible answers to this knowledge-type question are available, the question would have only been worth 4 marks and not 8 marks as for an understanding-type question, answered by Student A above.

Student B

Outline factors that should be assessed during a manual handling risk assessment (8 marks)

Load/Individual/Tile/Environment:

✓ 1 The load should be assessed for its weight (i.e. how heavy it is; it should not be too heavy for the person to lift).

✓ 2 Loads may be off-centre (i.e. if a box is loaded heavy on one side the load will be difficult to lift).

(1) 3 The load should be assessed for its weight (i.e. how heavy it is; it should not be too heavy for the person to lift).

4 Individual factors – the worker may be pregnant and

✓ therefore should not be lifting;

✓ nor should someone who has a bad back.

✗ 5 Tile?

✓ 6 Working environment: temperature, floors, lighting, etc.

(2) 7 Loads should not be unstable.

5/8

Student B did much better than Student A, scoring 5 out of 8 marks. Student B made notes (no marks available, but could be helpful in constructing an answer).

Reasons for marks awarded/not awarded are explained below:

1 **Assessed for its weight**
Correct: answer sheet has heavy, which is to do with the weight – therefore a mark may be given. Student B clearly outlined this particular point.

2 **Be off-centre**
Correct: mark awarded for unstable/unpredictable as referred to on answer sheet.

3 **Assessed for its weight**
Student B just repeated the first answer, so no mark given.

4 **The worker may be pregnant ... nor should someone who has a bad back**
Two marks can be awarded here. The above answer sheet has: hazard those with a health problem and hazard those who are pregnant. Even though Student B outlined these together, 2 marks would be awarded, as they are two different factors.

5 **Tile**
No marks: Student B when writing out his answer plan mistakes *Task* for *Tile*.

6 **Working environment temperature, floors, lighting, etc.**
Difficult to understand what Student B was trying to say here. Marks are available for an outline of poor floors; hot/cold/humid conditions or poor lighting conditions. However, this was not done, so only one mark can be awarded. If Student B had expanded on all of the three factors then an additional two marks would have been given.

7 **Loads should not be unstable**
Again this is a repeat of previous information given by Student B, only worded differently for answer number 2, so no marks can be awarded.

KEY NOTE

There is no need to underline your answer for the examiner, as this student has done. You must not use a highlight pen.

Student C

Outline factors that should be assessed during a manual handling risk assessment (8 marks)

Four factors are considered when carrying out a manual handling risk assessment:Task, Load, Individual and Environment.

Details may be found in the Manual Handling Operations Regulations 1992 (as amended), Guidance on Regulations,

HSE Document L23.

Under the factor of task the following questions may be asked:

✓ Does the task involve holding loads away from the trunk?

Will the employee be required to perform any of the following

✓✓ that would increase the likelihood of risk to them: twisting or stooping to pick up the loads.

✓ Reaching upwards should be avoided: having the heavier load stored

✓ at waist height. Likewise any large vertical movement should also be avoided.

✓ Will the work involve carrying loads over long distances?

✓ Will the work involve strenuous pushing or pulling?

The task should be carried out at a rate that the employee can operate

✓ safely (i.e. insufficient rest or recovery could cause long-term health problems).

✓ If a process imposes the work rate (i.e. the employee is forced to work harder than he or she is able), then this could lead to WRULD (work-related upper limb disorders).

Use of a MAC assessment for manual handling is advised by HSE.

8 max/8

Student C was able to gain maximum marks. Additional information was given which, even though it was correct and showed extended understanding of the topic, did not gain any additional marks. Here the student did not bother with an answer plan. Answer plans are not needed but some students do find them helpful.

Some of the extra information would more than likely gain marks in a diploma-level exam. The depth of understanding shown here is not normally examinable at award and certificate level:

- Details may be found in the Manual Handling Operations Regulations 1992 (as amended), Guidance on Regulations, HSE Document L23.
- Use of a MAC assessment for manual handling is advised by HSE.

Student D

Outline factors that should be assessed during a manual handling risk assessment (8 marks):

✗ Areas to be considered are the back, arms, legs, etc.

✗ Avoid slipped or prolapsed discs by wearing a lifting belt.

✗ Assess risk of strains, which are injuries that affect muscles, and sprains, which are injuries that affect ligaments.

✗ Reduce the risk of cuts and abrasions by wearing gloves when lifting sharp objects.

✓ If the worker is pregnant she should not lift anything as this might cause harm to the baby; her job role may have to change during the pregnancy. Under the Management of Health and Safety at Work Regulations a specific risk assessment should be carried out for pregnant workers. Employees are not to be discriminated against if they become pregnant. Pregnant employees must inform their employer.

1/8

Unfortunately, Student D did not answer the question correctly. The student answered this question from the angle of Avoid, Assess and Reduce manual handling tasks, which are the duties of the employer, and the injuries mentioned above are those that could result from poor manual handling. To gain marks you must answer the question set. One mark can be awarded for reference to the pregnant worker – the information provided by the student for the pregnant worker was extensive, but still only one mark can be given.

EXAMPLE 5: MISREADING MANUAL HANDLING

Outline factors that should be assessed for the load in relation to a manual handling risk assessment (8 marks):

L-I-T-E

✓ 1 The load should be assessed for its weight (i.e. how heavy it is) – it should not be too heavy for the person to lift.

✓ 2 Loads may be off-centre (i.e. if a box is loaded heavy on one side then the load will be difficult to lift).

✗ 3 If the worker is pregnant she should not carry out manual handling.

✗ 4 If an employee has a previous injury (bad back) this could increase the risk.

✗ 5 If the load has to be held away from the body this will increase the risk of injury – this should be prevented.

✗ 6 If the load has to be carried for long distances this will increase the risk of injury – using a trolley would reduce the risk.

✗ 7 Poor lighting can restrict vision and an employee could injure him/herself when carrying a load, as they may trip over something.

✗ 8 If the floor is uneven an employee could trip when carrying a load.

2/8

If the question was the same as for students A–D – **Outline** areas that should be assessed during a manual handling risk assessment – this would have scored 8 marks. Unfortunately, the wording of this question requires the student to **outline** only those factors to be considered with regard to the **load**. The student misread the question.

Outline factors that should be assessed for the Load in relation to a manual handling risk assessment (8 marks).

Only answers 1 and 2 related to the load. The answer plan of L-I-T-E sent the student off in the wrong direction. Answers 3 and 4 relate to the individual. Answers 5 and 6, even though they include the word 'load', are factors to consider relating to the task. Finally, answers 7 and 8 relate to the working environment. Therefore only 2 marks can be awarded.

KEYNOTE

Remember that all the awarding bodies have a bank of questions based on the learning outcomes for their respective courses. If you understand the learning outcomes, found in the syllabus of the course you are taking, you should be able to answer the questions to which they relate.

FINAL THOUGHTS

My hope is that this book will give you the essential study skills needed for you to successfully complete your course and to pursue a successful career, whether in health and safety or not.

Over the past 12 years I have had the great pleasure and privilege of teaching and consulting in the area of health and safety (in the UK and overseas); it started with taking the Professional Certificate Qualification – NEBOSH National General Certificate. I had no background in health and safety, and had been diagnosed with dyslexia, yet I was able to successfully pass this course. I say this to encourage you with the understanding that if I can gain such a qualification, so can you!

As you work through the principles of receiving, remembering and recalling the health and safety information from your course, and while you read this book, you will be developing transferable skills that may be used in the future for your work and your further studies.

The examples, case studies and key notes used in this book all follow a health and safety theme. The principles, however, outlined of the three Rs of Study – Receive, Remember, Recall – can in theory apply to any area of study.

If you have any questions or comments I would be very happy to hear from you. Please contact me via jonathanbackhouse@me.com

Finally, let me wish you all the best in your study and career.

APPENDIX
USEFUL BOOKS AND WEBSITES

BOOKS AVAILABLE FROM TAYLOR AND FRANCIS

Asbury, S. (2007) *Health & Safety, Environment and Quality Audits: A Risk-based Approach.*

Channing, J. (2014) *Safety at Work* (8th edn).

Hughes, P. and Ferrett, E. (2011) *Introduction to Health and Safety at Work* (5th edn). The Handbook for the National General Certificate.

Hughes, P. and Ferrett, E. (2011) *Introduction to Health and Safety at Work* (4th edn). The Handbook for the NEBOSH National Certificate in Construction: Health and Safety.

Hughes, P. and Ferrett, E. (2013) *Introduction to Health and Safety at Work* (2nd edn). The Handbook for the NEBOSH International General Certificate.

OTHER PUBLICATIONS

Armitage, A. et al. (2012) *Teaching and Training in Lifelong Learning* (4th edn). Berkshire: Open University Press.

Curzon, L.B. (2006) *Teaching in Further Education: An outline of principles and practice* (6th edn). London: Continuum.

Davis, R. (1997) *The Gift of Dyslexia.* Location: Souvenir Press.

Deveux, T. (2008) *Health and Safety for Managers Suprevisors and Safety Representatives.* CIEH, London.

Ott, P. (1997) *How to Manage and Detect Dyslexia.* London: Heinemann.

Reece, I. and Walker, S. (2007) *Teaching Training and Learning: A Practical Guide* (6th edn). Sunderland: Business Education Publishers.

Smythe, I. (2010) *Dyslexia in the Digital Age.* London: Continuum.

USEFUL WEBSITES

http://www.hse.gov.uk/pubns
https://www.britsafe.org
http://www.cieh.org
http://www.highfieldabc.com
http://www.rsph.org.uk
http://www.nebosh.org.uk

BIBLIOGRAPHY

In addition to accessing various syllabi from the awarding bodies, access to which is available via their respective websites, the following books where invaluable in developing this book.

Armitage, A. *et al.* (2012) *Teaching and Training in Lifelong Learning* (4th edn). Berkshire: Open University Press.

Backhouse, J. (2006) *Why Do People Lift Incorrectly Who Have Been Trained In Manual Handling?*, Sunderland University. Dissertation for BA, Sunderland University.

Backhouse, J. (2009) *Preparing to Teach – The First Steps.* Unpublished.

Backhouse, J. (2010) *The Impact of The Further Education Teachers' Qualifications (England) Regulations 2007 on Private Training Providers Delivering Health and Safety Training.* Dissertation for MA, Northumbria University.

Cohen, L., Mannion, L. and Morrison, K. (2004) *Research Methods in Education* (5th edn). London: Routledge Falmer.

Coombes, I. (2011) *A Study Book for the NEBOSH General Certificate: Essential Health and Safety Guide* (6th edn). Stourbridge: RMS Publishing.

Curzon, L.B. (2006) *Teaching in Further Education* (6th edn). London: Cassell Education.

Deveux, T. (2008) *Health and Safety for Managers Supervisors and Safety Representatives.* CIEH, London.

Fisher, R. (2001) *Teaching Thinking Philosophical Enquiry in the Classroom.* London: Continuum.

Graveling, R., Melrose, A.S. and Hanson, M.A. (2003) *The Principles of Good Manual Handling: Achieving a Consensus.* Edinburgh, Institute of Occupational Medicine.

Great Britain. Health and Safety at Work etc. Act 1974 (c 37). London: HMSO.

Health and Safety Executive (2004) *Manual Handling Operations Regulations 1992 (as amended). Guidance on Regulations* L23 (3rd edn). Norwich: HSE Books.

Health and Safety Executive (2006) *Essentials of Health and Safety at Work* (4th edn). Norwich: HSE Books.
Health and Safety Executive (2012) *Five Steps to Risk Assessment*. Available online at <www.hse.gov.uk/pubns/indg163.pdf>: HSE Books.
Hughes, P. and Ferrett, E. (2012) *Introduction to Health and Safety at Work* (5th edn). Abingdon: Routledge.
Reece, I. and Walker, S. (2007) *Teaching, Training and Learning* (6th edn). Sunderland: Business Education Publishers.
Reeve, P. (1995) *Passing the NEBOSH National General Certificate in Occupational Safety and Health*. The Tolleys Guide to Passing NEBOSH Exams.
Ridley, J. and Channing, J. (2008) *Safety at Work* (7th edn). Oxford: Elsevier.
Stranks, J. (2006) *Health and Safety Pocket Book*. Oxford: Elsevier.
Waugh, A. and Grant, A. (2001) *Ross and Wilson: Anatomy and Physiology in Health and Illness* (9th edn). London: Churchill Livingstone.

HEALTH AND SAFETY EXECUTIVE WEBSITE

Resources are available for all aspects of health and safety and may be found on the HSE website at http://www.hse.gov.uk/pubns/

Useful online resource from the HSE website include:

- Essentials of health and safety at work: The health and safety toolbox: How to control risks at work.
- HSG48 Reducing error and influencing behaviour.
- HSG65 Successful health and safety management.
- L21 Management of Health and Safety at Work Regulations 1999.
- L22 Safe use of work equipment. Provision and Use of Work Equipment Regulations 1998.
- L23 Manual handling. Manual Handling Operations Regulations 1992.
- L24 Workplace health, safety and welfare. Workplace (Health, Safety and Welfare) Regulations 1992.
- L25 Personal Protective Equipment at Work Regulations 1992.
- L26 Health and Safety (Display Screen Equipment) Regulations 1992.

The HSE book *Essentials of Health and Safety at Work* is now available online at http://www.hse.gov.uk/toolbox/index.htm

In addition, there are numerous industrial guidance documents (which will help you with your studies) also available (e.g. www.hse.gov.uk/pubns/indg143.pdf Getting to grips with manual handling).

INDEX